# Home How-To
# **Handbook**
## **Plumbing**

### **RICK PETERS**

**Sterling Publishing Co., Inc.**
NEW YORK

Library of Congress Cataloging-in-Publication Data Available

10  9  8  7  6  5  4  3  2  1

Published by Sterling Publishing Co., Inc.
387 Park Avenue South, New York, NY 10016

© 2006 by Sterling Publishing Co., Inc.

Distributed in Canada by Sterling Publishing
c/o Canadian Manda Group, 165 Dufferin Street,
Toronto, Ontario, Canada M6K 3H6
Distributed in the United Kingdom by GMC Distribution Services,
Castle Place, 166 High Street, Lewes, East Sussex, England BN7 1XU
Distributed in Australia by Capricorn Link (Australia) Pty. Ltd.,
P.O. Box 704, Windsor, NSW 2756, Australia

Sterling ISBN 13: 978–1–4027–4196–8
ISBN 10: 1–4027–4196–0

Book Design: Richard Oriolo
Photography: Christopher J. Vendetta
Page Layout: Sandy Freeman
Contributing Editor: Cheryl Romano
Copy Editor: Barbara McIntosh Webb
Indexer: Nan Badgett

For information about custom editions, special sales, premium and
corporate purchases, please contact Sterling Special Sales
Department at 800-805-5489 or specialsales@sterlingpub.com

# Contents

Introduction   4

**1** **Plumbing Basics**   6

**2** **Materials**   16

**3** Tools   42

**4** **Plumbing Know-How**   58

**5** **Installing Sinks**   92

**6** **Installing Faucets**   124

**7** Installing Fixtures   150

**8** **Troubleshooting**   180

Index   206

# Introduction

**W**HEN YOU THINK ABOUT IT, the idea of water flowing throughout your home is kind of scary. Water can quickly damage walls, cabinets, floors, and even the underlying structure of your home if it strays from its designated path. That's why it's so important that any work on your plumbing system be done with knowledge, skill, and quality parts.

Although it's great that so many plumbing fixture and fitting makers are producing a growing array of quality, do-it-yourself-friendly products and tools, that's only part of the equation. The other parts—knowledge and skill—are what this book is all about. We'll unravel the mysteries of home plumbing systems while helping you build the knowledge base and skills you'll need to tackle many plumbing projects and repairs.

## How to Use This Book

There are roughly three sections: Basics, Projects, and Troubleshooting. The first section (chapters 1–4) starts with *Plumbing Basics,* where we explore how the different parts of your plumbing system work. In *Materials* we guide you through the vast array of plumbing materials, parts, and products on the market.

*Tools* details the everyday and specialty tools you'll need to work on your plumbing system. In *Plumbing Know-How,* you'll find the nuts and bolts of working with pipe, fittings, and fixtures, making watertight connections, removing fixtures, and installing shut-off valves.

The second section, on Projects (chapters 5–7), includes *Installing Sinks:* how to install just about every kind of sink you'd want in your home. Then there's *Installing Faucets,* including centerset and widespread varieties. *Installing Fixtures* covers installing toilets, showers and shower doors, and water filter systems. The final section is *Troubleshooting,* where we share a systematic approach to tracking down and solving common plumbing problems.

## Codes and Permits

If any of your projects involve adding, extending, or modifying plumbing lines, check with your local building inspector for permit and inspection requirements. Usually, an inspector will first check your work at the "rough-in" stage (no wall coverings in place), and again when all the finish work is done (fixtures installed). By making sure your work is done to code, an inspector helps protect both your family and your home.

# 1

# Plumbing Basics

**P**LUMBING JOBS CAN BE STRAIGHTFORWARD—
or tricky. They can be neat—or really messy. Often
a single job can be all these things, but you can
minimize the tricky and messy parts. That's why before
you take on any plumbing job, you need to fully under-
stand how the three main systems in your home work
together: the supply system that provides fresh water
to all the fixtures in the home, the waste system that
removes liquid and solid waste, and the vent system that
lets the waste system drain properly.

# Supply System

Your supply system directs pressurized water to the plumbing fixtures throughout your home (fixtures are sinks, toilets, and tubs). Fresh water is supplied either by a water utility or from a private well; pressure comes from the city's pumps or from your well pump, respectively. Regardless of the source, water flows through a main shut-off valve (and a water meter if supplied by a utility), and then to the hot water heater. From there, both hot and cold water lines distribute water to various parts of the home. These lines are called branch lines. Lines that transport water up—to a second floor, or up into a fixture—are called risers.

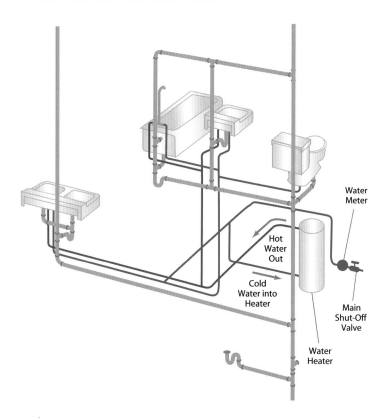

Water Meter

Hot Water Out

Cold Water into Heater

Main Shut-Off Valve

Water Heater

# Valves

**A**ll the hot and cold water lines in your home terminate in some sort of valve that, when opened, allows water to flow. Some valves are operated manually (like sink faucets); others are automatic (like those in toilets and refrigerator icemakers).

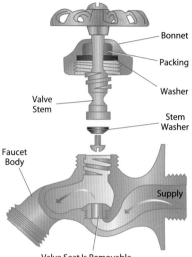

Bonnet

Packing

Washer

Valve Stem

Stem Washer

Faucet Body

Supply

Valve Seat Is Removable

## How a valve works

The simple valve in the drawing above is in the open position. The valve stem is up and water flows. When the handle is turned clockwise to the off position, the stem lowers until a rubber washer on the end of the stem meshes with a "seat" to halt the flow of water.

**PRO TIP**

## Shut-Off Valves

Every fixture in your home should have its own shut-off valve. If this isn't the case, consider spending part of a weekend installing one for every fixture (see pages 90–91). When an emergency occurs,

or it's simply time to make a repair, you (and your family) will appreciate that you can do the job without turning off the water for all, or part, of the house.

# Waste System

The waste system transports solid and liquid waste out of your home. The system uses gravity to move the wastewater from sinks, showers, tubs, and toilets out of the fixtures and into a waste line (often called the main or soil stack). This waste line empties into the municipal sewer or a private septic tank.

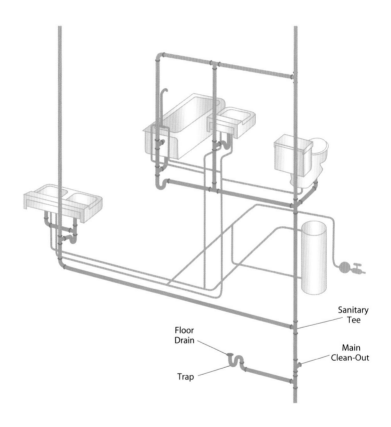

Sanitary Tee

Floor Drain

Main Clean-Out

Trap

# Traps and How They Work

In between every fixture and the waste line in your home is a trap—basically a curved section of pipe that captures or "traps" water. The trapped water forms an airtight seal to prevent sewer gas from entering the home. Since traps are curved, solid waste can build up in the trap and clog the line. Fortunately, most traps are easy to remove and clean out (see Chapter 8).

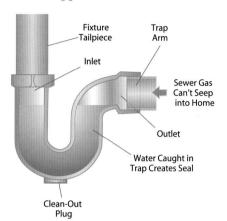

Fixture Tailpiece

Trap Arm

Inlet

Sewer Gas Can't Seep into Home

Outlet

Water Caught in Trap Creates Seal

Clean-Out Plug

## Types of traps

There are three basic traps: P-traps, S-traps, and integral traps. P-traps are the most common and are designed for waste lines coming out of a wall (instead of up through the floor as in an S-trap). Shaped like a P lying face down, P-traps are installed under sinks, tubs, and showers. S-traps were common when drain lines came up through the floor. But S-traps are prone to self-siphoning: The water seal can fail and allow sewer gas into the home. (S-traps are now prohibited in new homes.) Toilets have integral traps built into their bowls; just as with a P- or S-trap, a curved section traps water to create a seal.

P-Trap

S-Trap
(No Longer Legal)

Integral Trap
(Toilet)

# Vent System

The vent system is often the most unrecognized part of home plumbing. While it might seem that all a fixture needs to work is water and a drain, the third element, the vent, is critical for proper operation. Vents are connected along each fixture's drain line past the trap.

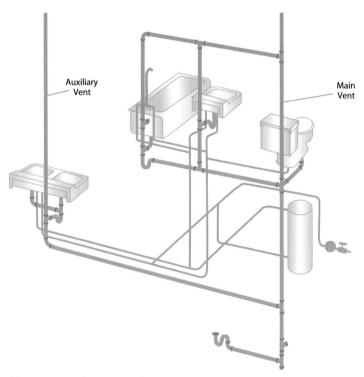

Auxiliary Vent

Main Vent

## How venting works

A vent does two important jobs: It lets wastewater in the drain system flow freely, and it prevents siphoning—the action that can pull the water out of traps, allowing sewer gas into the home. In both cases, a vent works by allowing fresh air to flow into the drain system (top drawing on page 13) the same way the second hole (or vent) in a gas can lets the gasoline flow out freely.

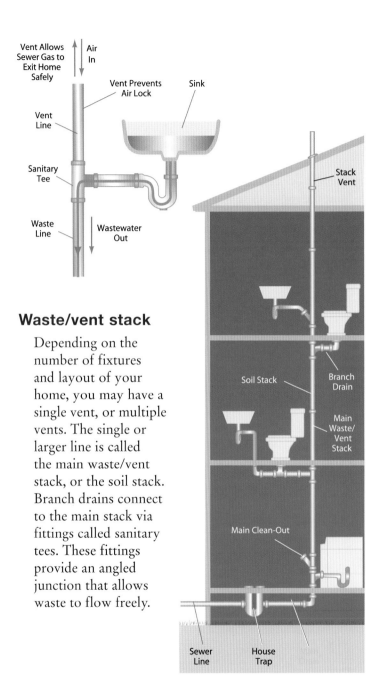

Vent Allows Sewer Gas to Exit Home Safely

Air In

Vent Prevents Air Lock

Sink

Vent Line

Sanitary Tee

Stack Vent

Waste Line

Wastewater Out

## Waste/vent stack

Depending on the number of fixtures and layout of your home, you may have a single vent, or multiple vents. The single or larger line is called the main waste/vent stack, or the soil stack. Branch drains connect to the main stack via fittings called sanitary tees. These fittings provide an angled junction that allows waste to flow freely.

Soil Stack

Branch Drain

Main Waste/ Vent Stack

Main Clean-Out

Sewer Line

House Trap

# Typical Bathroom

The typical bathroom has a sink, toilet, and bathtub or bathtub/shower combination. Hot and/or cold supply lines run to each fixture, each ending in a shut-off valve. Flexible supply lines connect the shut-off valves to the fixtures. The waste system carries away solid and liquid waste; traps prevent sewer gas from entering the home. Vents are connected along each fixture's drain line past the trap to allow fresh air in and sewer gas to flow safely out of the home.

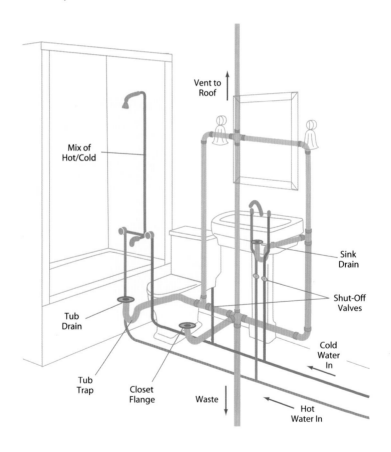

# Typical Kitchen

The average kitchen has a sink (with or without a garbage disposal), a dishwasher, and an icemaker hookup. Hot and/or cold supply lines run to each fixture. All lines terminate in a valve that, when opened, lets water flow. Valves such as sink faucets operate manually; other valves like those in refrigerator icemakers are automatic. The waste line relies on gravity to move the wastewater from sinks and dishwashers out into the municipal sewer or to a private septic tank. Vents connect to each fixture's drain line to allow fresh air in and keep sewer gas out of the home.

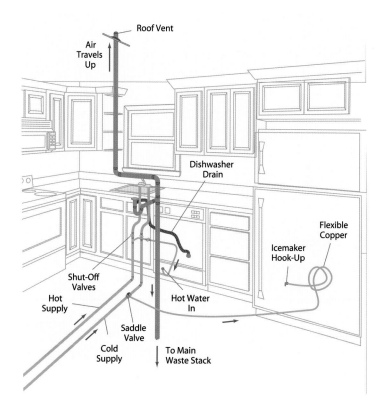

Roof Vent

Air
Travels
Up

Dishwasher
Drain

Flexible
Copper

Icemaker
Hook-Up

Shut-Off
Valves

Hot
Supply

Hot Water
In

Saddle
Valve

Cold
Supply

To Main
Waste Stack

# 2

# Materials

**M**UCH LIKE YOUR GROCERY STORE, there are aisles and aisles of choices in hardware stores and home centers today. But unlike, say, choosing the wrong pickles, selecting the wrong plumbing materials can mean a botched or delayed job— and possibly an emergency call to a plumber. You can sidestep these problems, though, with a basic understanding of the goods on the market, and their purpose.

# Pipe

Pipe allows water, waste, and air to flow throughout your home's plumbing system. The three most commonly used types of pipe are copper, PVC/CPVC (types of heavy-duty plastic), and galvanized metal. See the chart below for typical uses.

**MEASURING PIPE INSIDE DIAMETER.** In order to buy the correct pipe and fittings for a plumbing repair or remodeling job, you'll first need to identify the size of pipe you'll be working with. Pipe is sized based on the *inside* diam-

eter. If you have access to a cut section of the pipe, simply hold a tape measure or ruler across the widest part of the pipe. The inside diameter (ID), or nominal size, is the distance from one inner wall to the other.

## Supply and Waste Pipe Uses

| MATERIAL | USES |
|---|---|
| ABS | Drainpipes and traps |
| Cast Iron | Main drain waste |
| CPVC | Hot and cold water supply |
| Galvanized | Drainpipes; hot and cold supply lines |
| PVC | Drainpipes; vent lines |
| Brass | Valves and exposed traps and pipes |
| Chromed copper | Supply tubing for fixtures |
| Rigid copper | Hot and cold supply lines |

## Typical Pipe Sizes

| Copper OD | $3/8$" | $1/2$" | $5/8$" | $7/8$" | $1 1/8$" | $1 3/8$" | $1 5/8$" | |
|---|---|---|---|---|---|---|---|---|
| Copper ID | $1/4$" | $3/8$" | $1/2$" | $3/4$" | $1$" | $1 1/4$" | $1 1/2$" | |
| PVC OD | $7/8$" | $1 1/8$" | $1 3/8$" | $1 5/8$" | $1 7/8$" | $2 3/8$" | $3 3/8$" | $4 3/8$" |
| PVC ID | $1/2$" | $3/4$" | $1$" | $1 1/4$" | $1 1/2$" | $2$" | $3$" | $4$" |
| Galvanized OD | $5/8$" | $3/4$" | $1$" | $1 1/4$" | $1 1/2$" | $1 3/4$" | $2 1/4$" | |
| Galvanized ID | $3/8$" | $1/2$" | $3/4$" | $1$" | $1 1/4$" | $1 1/2$" | $2$" | |

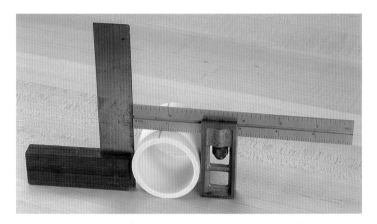

**MEASURING PIPE OUTSIDE DIAMETER.** In situations where you don't have access to a cut section of pipe, you can measure the outside diameter and use the chart above to determine the nominal ID of the pipe. One way to accurately measure the outside diameter (OD) is to wrap two squares (try, combination, or framing squares will work) around the pipe at right angles, as shown. Then read directly off one of the squares and check the chart for the inside diameter.

## Copper pipe

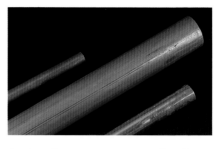

The supply lines that transport hot and cold water in most new homes are copper. Larger copper fittings and pipe can be used for small-diameter drainpipe, but more commonly plastic pipe is used for the waste lines.

**TYPICAL FITTINGS.** Connectors (called fittings, in the trade—not to be confused with faucets, etc.) make it easy to connect pipe into almost any configuration. They are sized for a specific-diameter pipe, usually ½" and ¾". Some common fittings are shown in the drawing below. A tee fitting lets lines intersect. Reducers join together pipe of different diameters. The 90- and 45-degree elbows turn corners. "Street" fittings are special elbows that have one male end and one female end; with this setup, you can join fittings together without using short sections of pipe (or nipples). A coupling joins two pieces of pipe and is commonly used in repair work for splicing together repaired sections of pipe. End caps seal off a pipe, whether for testing or for repair work. For step-by-step instructions on working with copper pipe and fittings, see pages 62–65.

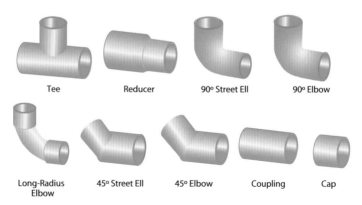

Tee    Reducer    90° Street Ell    90° Elbow

Long-Radius Elbow    45° Street Ell    45° Elbow    Coupling    Cap

## DWV pipe

Drain, waste, and vent (DWV) fittings come in white PVC (polyvinyl chloride) or black ABS (acrylonitrile-butadiene-styrene). (Note: Always match the fitting material to the type of pipe you're using—don't mix and match PVC with ABS—they use different cements.) DWV fittings have gentle curves instead of sharp corners, to keep wastewater flowing. Many fittings come in a "long-radius" bend for an even more gradual curve. Sanitary tees are used where waste lines converge. Clean-outs allow access for removing clogs. For step-by-step instructions on working with plastic pipe, see pages 66–68.

**CPVC PIPE.** Just like copper, CPVC (chlorinated polyvinyl chloride) pipe and fittings can be used for hot and cold supply systems. Make sure to check your local code to see whether CPVC is permitted in your area.

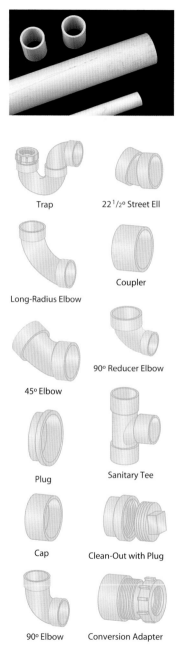

Trap

22$^1$/$_2$° Street Ell

Long-Radius Elbow

Coupler

45° Elbow

90° Reducer Elbow

Plug

Sanitary Tee

Cap

Clean-Out with Plug

90° Elbow

Conversion Adapter

## Galvanized pipe

Galvanized pipe is generally found only in older homes, where it's used as hot and cold supply lines. Although it's the strongest of all the types of pipe, it's also the most difficult to work with. Not only is it heavy, but it's prone to corrosion. Additionally, since the pipe and its fittings thread together, entire lengths often have to be removed in order to make a simple repair. For the most part, copper pipe (page 20) has replaced galvanized pipe, as it's much easier to work with and doesn't corrode. ABS pipe has replaced galvanized or cast iron pipe for waste lines: It's tremendously easier to work with and won't corrode. Common galvanized fittings are shown in the drawing below.

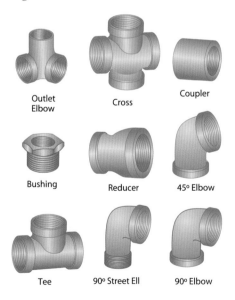

Outlet Elbow

Cross

Coupler

Bushing

Reducer

45° Elbow

Tee

90° Street Ell

90° Elbow

# Flexible Supply
# and Waste Lines

Flexible supply and waste lines are a homeowner's
dream come true. No specialty tools are required to
use them: with just an adjustable wrench or a pair of slip-
joint pliers, you're in business.

**FLEXIBLE
SUPPLY LINES.**
Flexible supply
lines typically
run between
shut-off valves
and fixtures—
usually in tight,
cramped spaces.

Any flexibility in these situations is appreciated. Flexible
supply lines are available with two different types of outer
protection: braided metal and vinyl mesh. They come in a
variety of pre-set lengths, complete with captive connecting
nuts. When buying flexible supply lines, you're better off
long than short; any excess can bend to one side. If they're
way too long, they can even be looped.

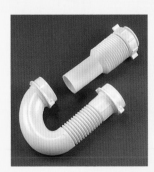

**FLEXIBLE WASTE LINES.**
Flexible waste lines connect
sinks and tubs to the waste
stack. They generally have an
accordion-style section that
can be bent and angled to
make the connection. Most
flexible waste lines use plastic
compression fittings to make
the connection.

MATERIALS

# Sealants

Sealants are the magic that keeps water flowing through your home where it's supposed to flow — and nowhere else. The four main sealants you'll use for your plumbing tasks are: plumber's putty, pipe joint compound, Teflon tape, and wax rings.

**PLUMBER'S PUTTY.**
Plumber's putty is a soft, dough-like substance available in cans or plastic tubs. It has been used for decades to create a watertight seal under the rims of sinks, under the base plates of faucets, and under the rims of sink strainers. It's inexpensive, is easy to work with, and does a great job. Although silicone caulk has become a popular alternative in many cases, most pros still stick with putty—they know it gets the job done.

**PIPE JOINT COMPOUND.** Many plumbing connections rely on threaded parts—flexible supply lines that connect to a faucet, shut-off valves that thread onto a copper fitting. The problem is that threads are not watertight. The original solution was to apply pipe

joint compound to the threads to create a water-tight seal. And pipe joint compound—commonly referred to as "pipe dope"—is still being used today; it's similar to tooth-paste and comes in squeeze tubes for ease of application.

**TEFLON TAPE.** The modern miracle of Teflon tape was the answer to many plumbers' prayers. Teflon tape is a thin membrane that wraps easily around pipe threads. And it creates a super-watertight seal—no mess, no fuss. Teflon

tape comes in small rolls and is quite inexpensive.

**WAX RINGS.** The final sealant is used only on toilets. A toilet bowl rests on a special DWV fitting called a closet flange. It's secured to the flange with closet flange or "Johnny" bolts. A wax ring—basically a donut-shaped chunk of wax—fits around the closet flange. When compressed, the ring creates a seal between the bottom of the bowl and the flange.

Q U I C K    F I X

# Graphite-Impregnated Cord

Older compression faucets use a packing washer instead of the more modern O-rings to create a watertight seal. You can still find packing washers at some hardware stores (if not, try a plumbing supply house), or you can wrap a few strands of graphite-impregnated cord around a faucet's stem. It's available at most home centers and hardware stores and can be used to stop larger leaks on almost any threaded part.

# Valves

**V**alves come in lots of shapes and styles, in either metal or plastic. Regardless of how they work internally, they all serve the same purpose—to open or close a water line. There are four main types of valves: ball valves, gate valves, saddle valves, and check valves.

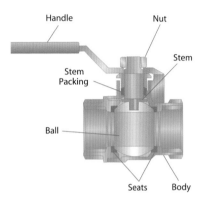

Handle     Nut

Stem

Stem Packing

Ball

Seats     Body

**BALL VALVE.** A ball valve can fully open or close with a quarter-turn of the handle. A "ball" inside the valve has a hole drilled through it to provide unrestricted water flow in the open position. Rotating the ball will position the hole in the ball 90 degrees to the water flow, effectively cutting off the water. Ball valves are becoming an increasingly popular choice for main shut-off valves in a home since they can turn water on or off in a hurry.

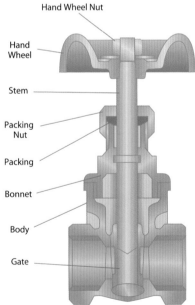

Hand Wheel Nut

Hand Wheel

Stem

Packing Nut

Packing

Bonnet

Body

Gate

**GATE VALVE.** Gate valves use a wedge-shaped brass "gate" to block off the flow of water. This type of valve is not as reliable as the ball valve described above, cannot be repaired if it doesn't fully stop water flow, and takes longer to turn water on or off. Note: For the most part, gate valves have replaced older globe valves (not shown) in new construction. Globe valves work similar to the way compression faucets do (see page 36).

**SADDLE VALVE.** Saddle valves are designed to "tap" into an existing supply line. A pair of base plates wrap around the existing line and are held in place with a set of machine screws. The valve itself is quite simple: It's a needle stem (with handle) and a rubber washer that fits in the bottom of the base plate, creating a water-tight seal. A small hole is drilled in the water line, and

the saddle valve is clamped in place. As the tapered tip of the needle stem is raised and lowered into the hole, water flows up into the valve and out its tee port.

O-Rings    Check    Body

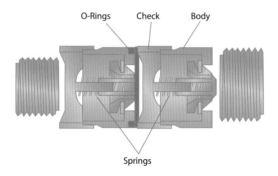

Springs

Double Check Valve

**CHECK VALVE.** Check valves are in-line self-activating safety valves designed to let water flow in one direction only. They may be single valves or double valves, as shown in the drawing above. Check valves are used in homes with wells to prevent damage to the pump by backflow from the pressurized tank.

# Bathroom Lavatories

A bathroom sink is often called a lavatory to differentiate it from a kitchen sink (page 30). Lavatories differ from kitchen sinks in size and bowl depth (drawing below). Unlike kitchen sinks, lavatories have an integral overflow to prevent water from flooding over the rim. To select a lavatory, consider first how it's mounted (page 29), what it's made from (page 32), and the size of the bowl. After that come choices in color and style.

**VANITY SINK.** Vanity-mounted sinks can be drop-in, under-counter, and above-counter.

**WALL-MOUNT SINK.** Any sink that is supported by a wall is considered a wall-mount sink. These sinks attach directly to the wall or hang on a bracket attached to the wall.

**PEDESTAL SINK.** Pedestal sinks may look freestanding, but the sink is mounted to the wall. Even though the pedestal takes some of the sink's weight, the bulk is supported by the wall.

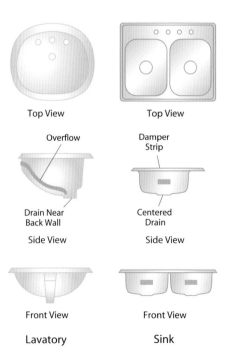

Top View     Top View

Overflow     Damper Strip

Drain Near Back Wall     Centered Drain

Side View     Side View

Front View     Front View

Lavatory     Sink

Sinks are often classified by how they are mounted: true self-rimming, tiled-in sink, self-rimming with clips, under-counter, and seamless; see the drawings below.

**TRUE SELF-RIMMING.** True self-rimming cast iron or porcelain sinks are held in place by their significant weight and a thin layer of sealant or plumber's putty that forms a watertight seal under the small flat section on the rim.

**TILED-IN SINK.** To have a sink that ends up flush with the countertop, you first install the sink and then build up the countertop around it, with tile.

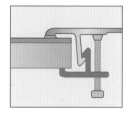

**SELF-RIMMING.** Although called self-rimming, this style of sink really isn't. To create a watertight seal, a dozen or so small clips hook onto a lip on the underside of the sink and pull the sink down tight.

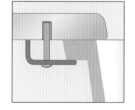

**UNDER-COUNTER.** Under-counter sinks have become popular since the advent of solid-surface materials. This style of sink presses up under the counter and is held in place with clips that screw into embedded inserts.

**INTEGRAL (SEAMLESS).** The ultimate solution to stopping water from leaking between a sink and its countertop is to form the sink and countertop as one unit. Or make the parts out of the same material, and glue the two parts together.

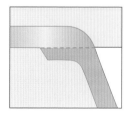

# Kitchen Sinks

Choosing a kitchen sink that will serve you well and provide the look and feel you're after takes some thought. In order of importance, you'll need to choose how the sink is mounted (page 29), what it's made from (page 32), the number and size of bowls, bowl depth, and the number of faucet holes.

**BOWL SIZE AND NUMBER.** Bowls vary from one to three; they can be identical in size and depth, or vary. Deeper-bowl sinks are becoming more popular, as well as sinks with a smaller bowl inset between two larger bowls. The smaller middle bowl frequently has a garbage disposer attached to it. Bowl number and size are really a matter of personal preference—if you pick a larger sink, just make sure it fits in your countertop.

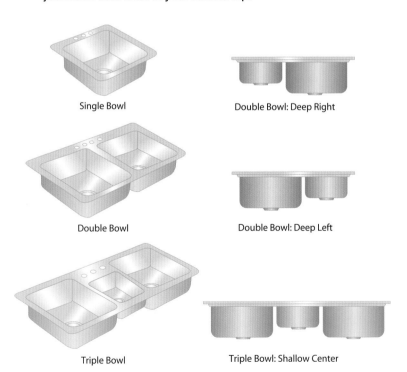

Single Bowl

Double Bowl: Deep Right

Double Bowl

Double Bowl: Deep Left

Triple Bowl

Triple Bowl: Shallow Center

**NUMBER OF FAUCET HOLES.** Finally, you'll need to match the number of faucet holes to the type of faucet you'll be using.

Three holes are the most common, as most faucets require at least this many holes.

Other sinks offer from one to five holes; the additional holes can be used for sprayers, filtered or hot water, or soap dispensers.

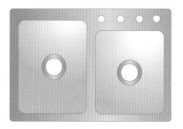

Offset

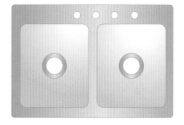

Centered

Centered

Single Hole with
Optional Holes

# Sink Materials

In the past, choosing a material for a sink was easy because you had only two choices: porcelain or cast iron. But today's sinks come in many materials, including china, cast iron, composites, metal, glass, and stone.

**VITREOUS CHINA.** A vitreous china sink is made of ceramic/porcelain that has been "vitrified" to create a glasslike surface that absorbs less water than most other ceramics. These are inexpensive, are easy to maintain, and come in the widest variety of shapes, sizes, and colors.

**CAST IRON.** Cast iron sinks aren't used much in new construction and remodeling anymore; they've been replaced by sexier composite and solid-surface sinks (see below). That's too bad, because cast iron is quiet, massive, and unmoving. The combination of the tough coating along with the heavy cast iron makes a formidable unit.

**COMPOSITE.** Composite sinks offer a big advantage over other sink materials—as long as they're being installed in a similar-material countertop. Their advantage is that they can be glued to the underside of the countertop to create a virtually seamless seam. Not only does this eliminate the possibility of leaks, but it also creates smooth-flowing lines and a rimless sink.

**STAINLESS STEEL.** For many years, stainless steel was seen as the ultimate kitchen sink material. Virtually free from staining, this hardy metal can be buffed out when scratched. Stainless steel is still an excellent option; it's relatively inexpensive, easy to find, and easy to install.

**GLASS.** Primarily the domain of high-end designers, glass sinks are available to the mass market through a bathroom distributor or designer. They are lovely, but it can take work to maintain their attractiveness.

**STONE AND ALTERNATIVE METALS.** Natural stone such as marble has been used for years for vanity tops. Lately, companies have started making sinks, too, out of other natural stones. Since these are often made to order, they are expensive, but their rugged natural surface holds up well for years. Non-stainless-steel sinks, such as the bronze sink shown here, are becoming popular—they look good and hold up well to wear and tear.

# Faucets

With all the options in faucets today, your best bet is to buy a name you can trust, and don't skimp on cost. A quality faucet will provide years of trouble-free service. But which one? Odds are that you've got an idea of the style you're after; but you'll also have to decide on control type, spout options, sprayer type (if applicable), and mounting options.

CONTROL TYPE. Your first step is to decide if you want separate controls or a single control. This is a matter of personal preference. Just keep in mind that the fewer the controls, the more holes you'll have available on the sink for other accessories such as a soap or hot water dispenser.

SPOUT OPTIONS. Another option to consider is spout style. The two most common styles are standard (bottom right) and gooseneck (bottom left). A goose-neck spout provides more room underneath the spout for washing hands and filling containers, but some folks find it may extend up too high and get in the way. This is a matter of personal choice.

**SPRAYER OPTIONS.** When it comes to choosing a sprayer, you can have it either separate (top left) or built-in (top right). A separate sprayer needs its own hole in the sink. Built-in or pull-out sprayers don't require this, so you can use the fourth hole in a sink for a hot or filtered water dispenser.

**MOUNTING OPTIONS.** There are two basic mounting options for faucets: centerset or widespread (see the drawing below). On a centerset faucet, the handles and spout have a common base plate. Supply lines connect directly to the valves in the handles. The handles are generally 4" on center. With a widespread faucet, the handles and spout are separate and although usually are mounted on 8" centers, can be mounted closer or farther apart. If the sink you'll be using is pre-drilled, this will determine which faucet type you can use. Generally, holes 4" apart will take a centerset faucet and holes 8" apart will accept a widespread faucet.

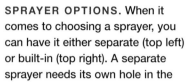

Hot $\frac{1}{2}$" NPT  Cold $\frac{1}{2}$" NPT
Lift Rod

Centerset

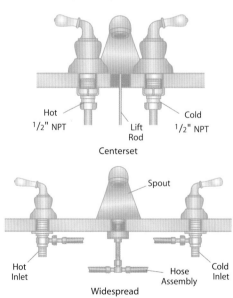

Spout

Hot Inlet  Hose Assembly  Cold Inlet

Widespread

## Faucet valve options

When choosing a faucet, it's helpful to know what type of valves the faucet uses to control the flow of water. The three main types form a study in faucet evolution, starting with compression valves, moving past the popular ball valve and on to the latest—and thought by many the best—cartridge valves.

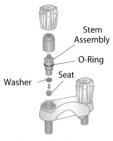

Compression Faucet

COMPRESSION. Compression faucets individually control the flow of either hot or cold water that is then sent to the spout. As the handle is rotated, a valve stem rises or lowers to allow more or less water through to the spout. A rubber washer on the end of the stem presses against the seat in the base of the faucet body to stop the flow when the handle is off.

BALL. The heart of a ball-type faucet is a ball that rotates as the single handle is pivoted. As the ball rotates, slots in the bottom align with the hot and cold water ports in the faucet body. A pair of spring-loaded seats press up against the ball to serve as simple on-off valves. When the slots align with the seat, the spring pushes up the seat, letting water flow.

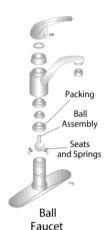

Ball Faucet

CARTRIDGE. In many ways, the cartridge-type faucet works just like a rotating ball faucet: As the single handle is pivoted, varying amounts of hot and cold water are directed up into the spout. The difference is how the water is directed. Instead of a ball that rotates against spring-loaded seats, a cartridge moves up and down within the faucet body to control the amount and temperature of the water directed to the spout.

Cartridge Faucet

# Toilets

If you haven't shopped for a toilet recently, you might be surprised to find varying capacities, sizes, and flushing actions. There are two styles of toilet: one-piece and two-piece (top drawing). The one-piece versions have a sleek look because the tank and bowl are, literally, one piece. Most toilets are two-piece, with a separate tank and bowl. The space you have for a toilet influences the bowl type (top drawing). A round-front toilet fits better in smaller spaces than an elongated bowl, which has more room in the front of the bowl.

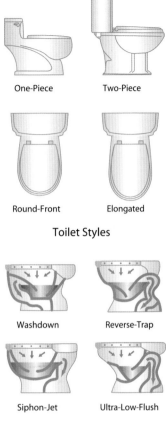

One-Piece     Two-Piece

Round-Front     Elongated

**Toilet Styles**

Washdown     Reverse-Trap

Siphon-Jet     Ultra-Low-Flush

**Flush Options**

**FLUSH OPTIONS.** Flush options include: washdown, reverse-trap, siphon-jet, and ultra-low-flush (middle drawing). Washdown traps are inexpensive but noisy; the bowl is flushed by streams of water draining from the rim. On a reverse-trap toilet, the bowl is longer but provides a quieter flush. With a siphon-jet toilet, a hole in the bottom sends a jet of water into the trap to create a siphoning action. This creates both a quiet flush and a large water surface area. An ultra-low-flush toilet has a low water table and correspondingly small surface area. This allows for a smaller tank and less water per flush; the drawback is that waste is often not completely cleared with a single flush.

# Bathtubs

**Y**our choice of bathtub types is generally defined by the existing space—unless you're remodeling. For the standard bath, a rectangular alcove tub (see below) is still the dominant (and least costly) choice.

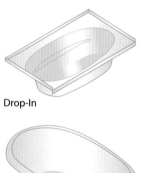

Drop-In

Freestanding

Corner

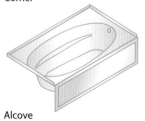

Alcove

**TYPES OF TUBS.** Bathtubs come in four main types: drop-in, free-standing, corner, and the most popular, alcove. Although drop-in tubs can be installed in an alcove, they are typically dropped into a stand-alone base frame. Freestanding tubs can either rest on the floor or be raised up on legs (like the old claw-foot tubs of days past). A corner-styled rectangular tub will have a finished apron, and one or more finished ends. A recessed or three-wall alcove tub is surrounded on three sides by walls, and has a finished front section or apron.

**MATERIALS OPTIONS.** While several materials compete for your tub investment, the best bets come down to two: enameled cast iron, very durable and very heavy, and acrylic (including solid-surface versions), which is both durable and lightweight. A maintenance advantage with acrylics is that the color is solid throughout, so visible scratching and wearing are minimized.

# Showers

Today's shower designs, materials, and colors offer choices like never before. Component pieces let you make separate purchases of the surrounds (walls), the shower bases (also called shower pans or receptors), and the doors; see the drawing below. Showers are also available as one-piece units. Although it may seem like the easiest way to create a watertight surround, one-piece (also called modular) surrounds do have a major drawback: their size. All too often, homeowners bring one home, only to find that they can't get it into their house or their bathroom. This type of surround is best used in new construction, where they can be installed before the doors and wall coverings are in place.

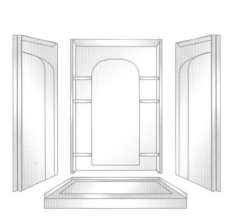

Multi-Segment

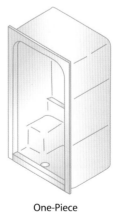

One-Piece

# Shower Doors

Tired of grappling with a shower curtain? Step up to a shower door. You'll find a wide variety of easy-to-install units available. Options include metal finish (chrome, brass, and brushed metal are common), along with door type and glass options.

**DOOR OPTIONS.** Most shower doors are sliding (often called bypass doors, in the trade); see the drawing below. Both doors may slide, or one may be permanently fixed in place. These doors are the most economical in terms of space. If your bath is roomy, you may prefer a pivoting door. These provide better access in and out of the shower, but require clearance for the door to swing.

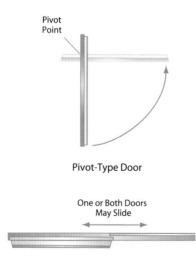

Pivot Point

Pivot-Type Door

One or Both Doors
May Slide

Sliding or Bypass Door

**GLASS OPTIONS.** Glass options to choose from include clear (top photo) or frosted glass (middle photo), along with added choices in patterns like fluted, pebbled, wavy, or smooth.

# Luxury Fixtures

Two of the more common luxury upgrades for bathrooms are jetted tubs and spa showers.

JETTED TUBS. A jetted tub (often called a whirlpool) takes water from the tub, mixes it with air, and pumps it back through jets that may adjust for pressure and direction. You can find jetted tubs in many shapes—round, oval, or rectangular—or in three-corner models. Whatever the shape, they all have special requirements. The pump needs a separate ground-fault-protected electrical line and must be accessible for servicing. Since a jetted tub can be much heavier than a conventional bathtub, it may require extra floor support, too; check with a professional if you're unsure.

SPA SHOWERS. With a spa showers like the one below, taking a shower can become a highly customized experience. This unit's built-in transfer valve lets you direct the water from the shower-head alone, from the body sprays and handheld shower, or from all three sources simultaneously. Because the body sprays are adjustable, each user can adapt the unit to personal preferences. If you're thinking about upgrading to a spa shower, it's best to hire a professional—the plumbing required to feed the multiple spray outlets can be quite complex (pressure-balancing loops, etc.), and that's a job for experienced hands.

MATERIALS

# 3

# Tools

**M**ANY OF THE TOOLS YOU'LL NEED for your plumbing projects are everyday items, like wrenches and screwdrivers. But some of the tools—like a basin wrench or a closet auger—are quite specialized and in many cases, the job just can't be done without them. In this chapter, we'll describe all of the essential tools you'll need to keep water flowing where it should, and not where it shouldn't.

# Wrenches

At the heart of a plumber's toolbox are various gripping tools for loosening and tightening nuts, caps, fittings, valves stems, etc. These include many types of wrenches: pipe, adjustable, sliding-jaw, locking-jaw, open-ended, box, and offset.

**PIPE WRENCH.** An icon in days past, the pipe wrench was the plumber's main tool, since pipe was either galvanized or cast iron. The deeply serrated jaws of the wrench dig into the metal to grip the pipe. But the advent of copper and plastic pipe has caused a decline in the need for this wrench.

**ADJUSTABLE WRENCH.** Often known by the brand name Crescent, an adjustable wrench has one fixed and one adjustable jaw. When you turn the captive worm screw, it forces the rack on the adjustable jaw to slide open or closed. Because the jaws aren't serrated, an adjustable wrench is a good choice when tightening or loosening exposed parts like the chrome nuts on shut-off valves.

**SLIDING-JAW WRENCH.** On a sliding-jaw wrench (commonly known by the brand name Channellock), the width of the jaw opening depends on tongues on one half of the tool locating in grooves or channels in the other half of the tool. Since the jaws are fully adjustable and serrated, a sliding-jaw wrench can handle a wide range of connectors.

**LOCKING-JAW WRENCH OR PLIERS.** Better known by its brand name Vise-Grips, a locking-jaw wrench is an invaluable plumber's tool. The opening of the serrated jaws is controlled by turning a knurled knob on the end of the wrench. But what's special about this wrench is that the jaws will remain closed or "locked" around a part until a release lever is oper-ated. This makes it an ideal "third hand".

**OPEN-ENDED AND BOX WRENCHES.** Box wrenches—where all sides of the wrench's jaws are enclosed (bottom in photo)—were originally developed for square nuts and bolts. When hexagonal heads were introduced, one end of the "box" was left open (top in photo) so the wrench could slip onto the hex head easily. Nowadays wrenches often come with one open end and one box end (middle in photo).

**PRO TIP**

## Offset Wrenches

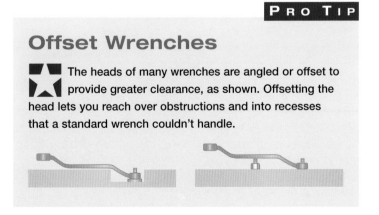

The heads of many wrenches are angled or offset to provide greater clearance, as shown. Offsetting the head lets you reach over obstructions and into recesses that a standard wrench couldn't handle.

# Pipe-Cutting Tools

For many plumbing tasks, you need to cut and fit pipe—either plastic or copper. The cutting tools to have handy are a hacksaw and mini-hacksaw, tubing cutters, and plastic-tubing pliers.

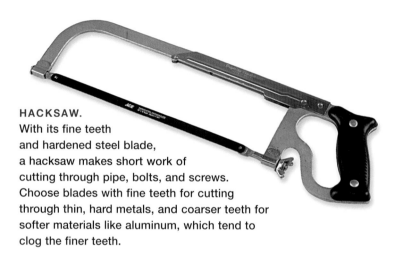

**HACKSAW.**
With its fine teeth
and hardened steel blade,
a hacksaw makes short work of
cutting through pipe, bolts, and screws.
Choose blades with fine teeth for cutting
through thin, hard metals, and coarser teeth for
softer materials like aluminum, which tend to
clog the finer teeth.

**MINI-HACKSAW.**
This small cousin of the
hacksaw uses the same-sized blade
but in a much smaller frame. The frame
is a combination handle/blade holder and will
let you get into places that a regular hacksaw
can't access.

**TUBING CUTTER.** The most common tool for cutting copper pipe is the tubing cutter. A small but very sharp wheel is pressed into the pipe, and the pipe or cutter is rotated. Tightening the wheel, coupled with continued rotation, results in cleanly cut pipe (for more on using a tubing cutter, see page 62).

**MINI-TUBING CUTTER.** A mini-tubing cutter is just a smaller version of its big brother. Just as with a mini-hacksaw, a mini-tubing cutter lets you get the job done in tight quarters.

**PLASTIC-TUBING PLIERS.** Plastic-tubing pliers have sharp blades attached to a fixed jaw and a ratcheting jaw to cut cleanly through small-diameter plastic pipe. Larger plastic pipe can be cut with a handsaw, but more accurate cuts are easily made with a power miter saw (see page 50).

TOOLS

# Copper-Pipe Tools

I f you plan to join copper pipe (commonly referred to as "sweating pipe," in the trade), you'll need some special tools, including a reamer, wire-brush pipe cleaners, a propane torch and sparker, flux and a flux brush, and heat shields. Note: Since joining copper pipe (see page 65) uses molten solder to join or "sweat" the pipes together, you must protect yourself. Make sure to wear safety glasses when working around open flames and heated parts.

**REAMER.** Although most tubing cutters (see page 47) come with built-in reamers for removing burrs in the ends of cut pipe, a reamer with a pivoting tip will do a better job (see page 63 for how to use this nifty tool).

**WIRE-BRUSH PIPE CLEANER.** Copper oxidizes when exposed to air. The pipe and fittings you buy will have oxidized, and you'll need to remove this oxidation to get a good solder joint. The best tool for this is a combination wire brush—it has two sizes of spiral brushes on its ends to clean inside ½" and ¾" fittings; two wire brush holes in the handle make cleaning the ends of ½" and ¾" pipe a snap.

**TORCH AND SPARKER.** The heat for sweating pipe comes from a propane torch. You can buy the torch head and propane canister separately or as a kit. Kits usually include a sparker for igniting the torch and a snap-on fan tip for spreading the flame (these are great for thawing frozen pipe).

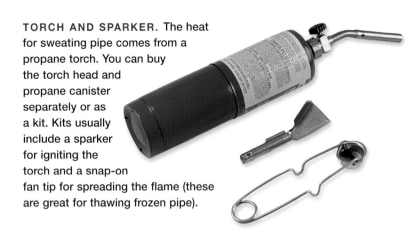

**FLUX AND BRUSH.** Think of flux as a "pipe sweating" helper. When applied to both fittings and pipe and then heated, flux paste (looks like a cross between candle wax and petroleum jelly) both cleans the joint and helps the solder to flow. You apply flux with a short-haired disposable flux brush.

## HEAT SHIELDS

On some projects you'll need to sweat pipe in a confined space, such as between studs when installing a shower valve. To prevent the heat of the torch from igniting nearby walls and framing, use a heat shield. The two most common types are shown here: One is a woven fire-retardant mat; the other is a spun heat-resistant material attached to a thin metal cover to help dissipate heat. Note that neither of these is fire-proof.

# Plastic-Pipe Tools

**A**lthough plastic pipe can be worked with ordinary hand tools, the job will go faster with specialized tools like a deburring tool, primer and cement, and a power miter saw. This is especially true if you're installing a lot of pipe.

**DEBURRING TOOL.** When you cut plastic pipe, you'll end up with burrs. To promote the flow of waste and water, these burrs must be removed. You can use a utility knife (page 52), but a pipe reamer like the one shown here does the job quickly and cleanly (page 67).

**PRIMER AND CEMENT.** The beauty of plastic pipe is that it's joined with cement, not solder. Depending on the pipe, you may need to prime it before applying cement. Most primers have a built-in dauber. It's important to note that all plastic-pipe cement isn't the same—you'll need to buy cement for the specific material you're joining (see page 68 for more on joining plastic pipe).

**PRO TIP**

## Miter Saws

Equipped with a carbide-tooth saw blade, a power miter saw will slice through even the thickest plastic pipe like butter. Power miter saws are particularly handy when you need to trim off just a bit— 1/8" or so—something that's quite difficult with a handsaw or tubing cutter.

# Remodel Tools

Some of the plumbing jobs you'll tackle will require some remodeling, like tearing out an old section of wall, flooring, or a cabinet. You'll find these tools useful: an electronic stud finder, a reciprocating saw, and a set of hole saws.

ELECTRONIC STUD FINDER. If any of your plumbing projects require cutting into a wall, you'll first need to locate the wall studs. An electronic stud finder is the tool for the job. Recent advancements in technology have driven down the price of these tools while jacking up their accuracy.

RECIPROCATING SAW. A reciprocating saw (usually known by the brand name Sawzall) can make quick work of cutting through wall coverings and framing.

HOLE SAWS. Whenever pipe passes through framing, flooring, or walls, you'll need to cut an access hole for the pipe. The most accurate way to do this is with a hole saw. A hole saw has a drill bit centered on a cup-shaped saw blade. With a set, the drill bit (or pilot) is inter-changeable between blades.

# Hand Tools

In addition to the specialty tools described in this chapter, your toolbox should include these essential tools: a tape measure and level, utility and putty knives, screwdrivers, a socket set, a hammer, and an electric drill.

**LAYOUT TOOLS.** One of the most critical steps in any plumbing remodeling job is measuring and laying out the placement of piping and fixtures. Two tools are the most handy for this: an accurate tape measure and a small torpedo level.

**UTILITY AND PUTTY KNIVES.** For plumbing work, you'll find a utility knife great for deburring pipe, cutting drywall, and trimming gaskets. A putty knife is useful for removing old fixtures and packing putty in faucet bases.

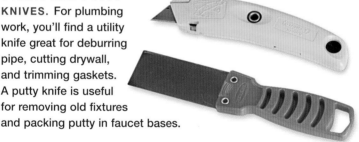

**SCREWDRIVERS.** Multi-tip screwdrivers (like the top screwdriver in the photo) pack a lot of punch in a small space. At minimum, you'll want a 4-in-1 that features two different-sized Phillips and slotted bits. An offset screwdriver can get into places a standard screwdriver can't.

**SOCKET SET.** A ratchet with interchangeable sockets is invaluable when working with the various nuts and bolts involved with plumbing projects. In addition to both imperial and metric sockets, you'll find a socket extension a real knuckle-saver.

**CLAW HAMMER.** Whether used for knocking out notches in studs, attaching shower flanges to wall studs, or ripping out or installing new framing, a standard claw hammer is a must in every plumber's toolbox.

**ELECTRIC DRILLS.** The two types of electric drill you'll find most useful for plumbing work are a cordless drill and a right-angle drill. Since much plumbing work is done in tight quarters (such as between wall studs), a right angle-drill lets you access those hard-to-reach places. Note how the drill chuck on a standard drill (right in photo) is in line with the motor, versus the chuck on a right-angle drill (left in photo), which is at 90 degrees to the motor.

TOOLS

# Waste-Line Tools

When a clog creates a plumbing emergency, reach first for a plunger. Then, if necessary, use a closet auger, and lastly, a snake or hand auger.

**CLOSET AUGER.** A closet auger is a special type of snake or auger (see page 55) designed for use only on toilets. Instead of forcing the clog down the toilet's trap, a closet auger is designed to retrieve objects stuck in a toilet's trap.

PRO TIP

## Plunger Options

Many clogs can be cleared with a plunger. It's a good idea to keep a few types on hand, as shown in the drawing at right. Either a flanged (force cup) plunger or a closet bowl force cup is your best bet for clearing a toilet clog (see page 190 for more on this). A non-flanged or sink plunger works best for flat-bottom sinks, like a standard kitchen sink.

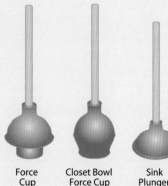

| Force Cup | Closet Bowl Force Cup | Sink Plunger |

**SNAKE.** A drain snake is just a loop of tightly coiled wire with a larger loop on its end that is forced down a drain. Most snakes come with an adjustable handle that slides down the coil and can be locked in place. Once the handle is locked, you twist it to force the tip of the snake to burrow into the clog. Then the clog can be pulled out or forced down the drain.

**HAND AUGER.** A less messy version of the snake is a hand-cranked auger. Here, the snake is coiled within a case and pulled out as needed. Once the clog is encountered in the drain, the snake is locked in place. Turning the knob on the back of the case forces it to rotate and dig into the clog. (For more on using snakes and augers, see Chapter 8.)

see Chapter 8.

## P R O  T I P

# Why Pros Don't Use Chemical Clog Removers

Clogs in sinks are usually caused by a buildup of soap, hair, toothpaste, food residue, etc. Although a common knee-jerk reaction is to reach for a chemical drain cleaner—don't. Most commercial brands aren't strong enough to clear clogs; professional-strength versions do a better job, but there's risk involved: The chemical reaction that clears the clog can also generate sufficient heat to damage your pipes, particularly plastic traps and clean-outs. Instead of chemicals, pros will always reach for a plunger. If the problem persists, they'll remove the trap to clear debris; or if that doesn't work, they'll use a snake or auger.

# Specialty Tools

Sometimes there's one and only one tool for the job, and nothing else will do. In plumbing, these tools are: a basin wrench, a strap wrench, a strainer wrench, a spud wrench, a seat wrench, and a tube-flaring tool.

**BASIN WRENCH.** Even in tight spaces, a basin wrench lets you reach up and loosen nuts. This wrench features a long extension arm for loosening or tightening faucet-mounting nuts below a sink. A standard wrench is hard to adjust in a limited-access space. The jaws of a basin wrench, though, are self-adjusting: As you apply pressure, the jaws close to fit around the nut.

**STRAP WRENCH.** If your plumbing project will involve working with large-diameter pipe, like the main waste/vent stack in your home, you'll find that a strap wrench is the only wrench that can handle the job. In use, you pass the wrench's strap around the pipe and back into the handle, where it's pulled tight as you exert pressure on the handle.

**STRAINER WRENCH.** Although you can remove or install a strainer with a pair of needle-nose pliers, that method tends to damage the strainer. A less harmful alternative is to use a strainer wrench. One end sports three tabs, the other four. This way, the tool will fit in strainer baskets with either three or four ribs. A metal cross dowel makes it easy to rotate the wrench.

**SPUD WRENCH.** No, they're not for taters— the spud wrench was originally designed to handle the large spud nut on the bottom of a toilet tank. Today's spud wrenches come with fixed or adjustable jaws and can handle a variety of large nuts. Since the jaws aren't serrated, they can loosen and tighten the large slip nuts found on metal sink traps without damaging the nut.

**SEAT WRENCH.** A seat wrench has nothing to do with toilet seats: Its sole purpose is to remove and install the seats inside a compression faucet (see page 187). The seat in a quality compression faucet is replaceable, and the L-shaped seat wrench is the only way to access the seat. Different tips on the wrench fit different types of seats.

**TUBE-FLARING TOOL.** If you'll be working with flared fittings (see page 73), you'll need this special tool to flare the pipe end out like a small trumpet. This two-piece tool consists of a clamp with various-sized holes to hold different-diameter pipe, and a T-handled flaring tool that has a tapered tip to create the flare on the end of the pipe.

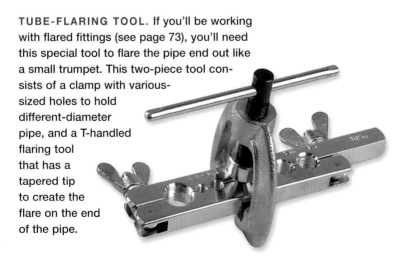

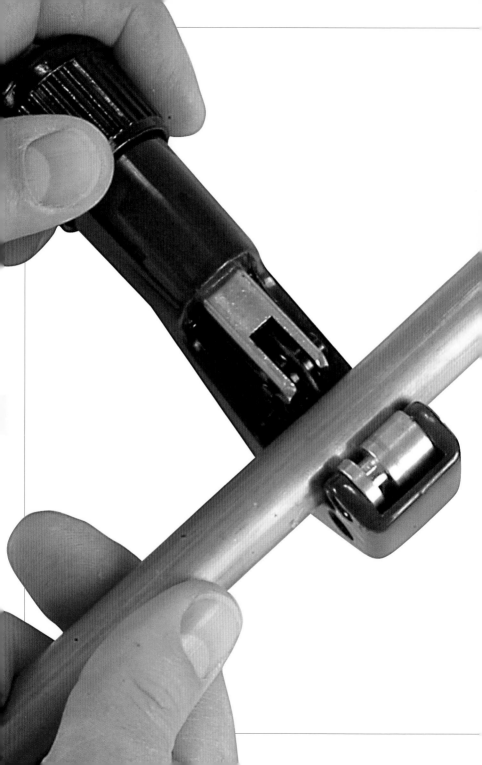

# 4

# Plumbing Know-How

**W**ITH SOME HOME IMPROVEMENT JOBS, like painting a room, a goof is easily fixed: You just repaint. Plumbing slipups, though, can have really nasty consequences—you could have water pouring out of the walls. That's why it's so important to have a solid grasp of the fundamentals, like joining pipe to create watertight joints. We've packed this chapter with the essentials you'll use on every plumbing project.

# Working with Wrenches

Virtually every plumbing project you'll tackle will involve working with wrenches—adjustable wrenches, sliding-jaw wrenches, locking-jaw wrenches, and occasionally a pipe wrench.

**TIGHTENING AND LOOSENING.** If you haven't worked with wrenches a lot, it's easy to get confused about which way to turn the wrench to tighten or loosen a nut or bolt. For around 98% of the nuts and

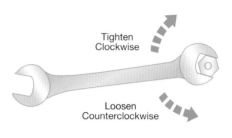

Tighten Clockwise

Loosen Counterclockwise

**Top View**

bolts you'll be dealing with, you'll move the wrench clockwise to tighten and counterclockwise to loosen, as shown in the drawing above. The other 2% of nuts and bolts are *reverse*-threaded—you'll find these used to hold spinning parts, like wheels, in places where the natural rotation of the wheel would loosen the nut or bolt as it spins.

**TWO-WRENCH METHOD.** Many plumbing tasks involve using two wrenches together to tighten or loosen a part. If space allows, use a wrench in each hand. For tight quarters, use just one hand. Place one wrench on one part

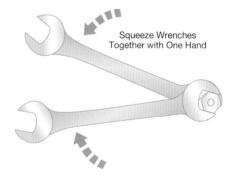

Squeeze Wrenches Together with One Hand

and the other wrench on the other part, so the wrenches are splayed slightly, as shown in the drawing above. Then squeeze to loosen or tighten. The pressure from one hand is often all you'll need to tighten or loosen a part.

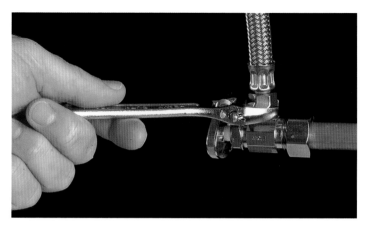

**ADJUSTABLE-JAW WRENCHES FOR FINISHED PARTS.**
Many of the plumbing parts you'll be working with will be exposed, and will have a decorative finish such as chrome or brass. If you use any wrench that has serrated jaws to tighten or loosen the part, the jaws will cut into and damage the finish. Whenever possible, use an adjustable wrench or other unserrated-jaw wrench to tighten or loosen the part; the smooth jaws won't hurt the finish.

**PRO TIP**

# Preventing Damage from Serrated Jaws

To tighten or loosen decorative parts that are too large for an adjustable wrench, wrap a couple turns of duct tape around the serrated jaws of a sliding-jaw wrench to pad the jaws and prevent damaging the finish.

# Working with Copper Pipe

**C**opper pipe is easy to cut and is joined together using a variety of fittings (page 20). A watertight seal is created by soldering—called "sweating," in the trade—parts together with molten solder. Although this may sound dicey, it's really quite simple once you've done it a few times. That's why it's a good idea to practice first on scraps before working on an actual project.

## Cutting pipe

There are two ways to cut copper pipe: with a tubing cutter and with a hacksaw.

**WITH A TUBING CUTTER. To** use a tubing cutter, open its jaws until it can be slipped over the pipe. Position it on the pipe so the cutting wheel is directly over a mark for the desired length. Tighten the knob and rotate the tubing cutter around the pipe; continue alternately tightening the knob and rotating the tubing cutter until you've cut through the pipe.

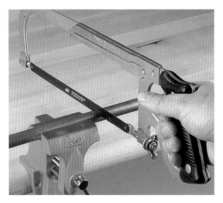

**WITH A HACKSAW.** You can also cut copper pipe with a hacksaw. The disadvantage to using a hacksaw is that the cut end will need filing to remove burrs, and it can be difficult to cut the end square—something a tubing cutter does automatically.

## Deburring

The cutting action of a tubing cutter creates a lip (basically a continuous burr) all the way around the inside of the pipe. This lip will restrict water flow and needs to be removed—you can do this with the built-in reamer on a tubing cutter or with a hand-held pivoting reamer.

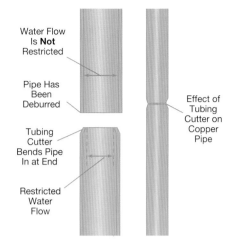

Water Flow Is **Not** Restricted

Pipe Has Been Deburred

Tubing Cutter Bends Pipe In at End

Restricted Water Flow

Effect of Tubing Cutter on Copper Pipe

**WITH A TUBING CUTTER.** A couple of twists will do the job. Most tubing cutters have a built-in tapered reamer that pivots out over one end. Extend the reamer and insert it in the cut end of the pipe. Rotate the reamer or the pipe to shave off the internal burr.

**WITH A REAMER.** The problem with reamers built into tubing cutters is that they're not very efficient and tend to dull easily. A better tool for the job is a handheld reamer. Just insert the pivoting tip into the pipe end and rotate the pipe to shave off the burr.

## Preparing to sweat pipe

Copper oxidizes when exposed to air. So all your pipe and fittings will be oxidized, and oxidation is the nemesis of a good solder joint. Solder needs fresh copper to flow and harden into a watertight joint—that's why you need to clean both the pipe and fittings that you'll be joining.

**CLEANING PIPE.** You can clean pipe in two ways. The easiest is to use a wire brush tool, found in the plumbing supply aisle. Insert the pipe into the tool and rotate until the pipe end is as bright as a freshly minted penny. Alternatively, wrap a strip of emery cloth around the pipe and rotate to clean.

**CLEANING FITTINGS.** To remove oxidation from the inside of fittings (the outside is irrelevant), use a spiral wire brush. Insert the brush in the fitting, and twist or rotate until clean.

PRO TIP

## Marking Alignment

It's always a good idea to make an alignment mark across a pipe and a fitting so you can realign the parts after they've been cleaned and fluxed. Unfortunately, the heat from a torch will burn off most marks. That's why plumbing pros (and blacksmiths) use soapstone—it's impervious to heat. Soapstone pens can be found in the plumbing supply aisle.

## Sweating copper pipe

There are three steps in sweating copper pipe: apply flux, heat, and apply solder.

**1** **APPLY FLUX.** Apply flux to the pipe end and inside the fitting with a flux brush. When heat is applied, the flux will burn off any remaining oxidation and also help the solder flow into the joint. Apply a generous dollop to both the pipe and the fitting with a flux brush. Push the flux-covered parts together until they bottom out. If you've made alignment marks, twist the pipe until they align.

**2** **HEAT THE PARTS.** Light your propane torch and apply the flame to the pipe and fitting. The more metal you're working with, the longer it will take to heat up. Move the flame back and forth in gentle sweeping motions, concentrating on the fitting. The flux will begin to sputter and spurt as it melts and cleans the joint. If you're working on existing plumbing, protect nearby surfaces with a heat shield (page 49).

**3** **APPLY SOLDER.** Withdraw the heat, and touch solder to the joint. If it melts, it's ready to go. If it sticks to the fitting, apply more heat and try again. Continue applying solder to the joint—it will wick into the fitting until it's full. You should see a continuous bead of molten solder around the entire joint.

# Working with Plastic Pipe

**B**y far the simplest pipe to work with, plastic pipe can be cut and joined easily. Just remember that all plastic pipe is not the same (see page 21), so make sure you're joining like materials (PVC to PVC). Dissimilar materials are not compatible and will degrade over time.

**1** **CUTTING PLASTIC PIPE.** There are a number of ways to cut plastic pipe: with plastic tubing pliers, with a hacksaw, or with a power miter saw (see page 50). The important thing is to make sure the cut is square. If it isn't, the joint will be weak and the seal questionable at best.

**WITH PLASTIC TUBING PLIERS.** Open the knife jaws of the pliers and insert the pipe. Squeezing the handles together will cause the jaws to ratchet closed; continue until you've cleanly cut through the pipe.

**WITH A HACKSAW.** If you don't have a lot of pipe to cut, you can get by just fine with a hacksaw. Clamp the pipe in a vise and cut through, taking care to keep the cut end square.

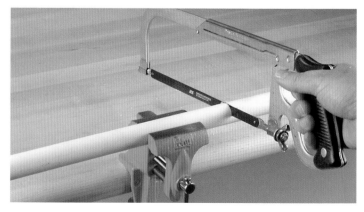

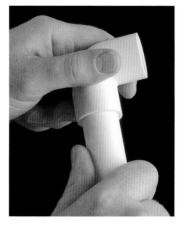

**2 DEBURR THE ENDS.** However you cut your pipe, you'll need to clean up the ends. You can use a nifty plastic-pipe deburring tool for this (page 50), or run a utility knife around the inside edge to remove any burrs, which could clog up an aerator or worse, restrict water flow.

**3 TEST THE FIT.** With your pipe cut and deburred, take the time to reassemble all your parts to make sure everything fits. Testing the fit of your cut pipe and fittings can save you the trouble and expense of redoing a job. Once cement is applied, it sets almost immediately.

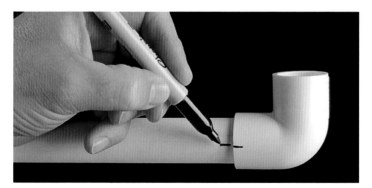

**4 MARK ALIGNMENT.** After you've assembled your pipe in place, draw an alignment mark across each joint. This way you can take the pieces apart, apply cement, and reassemble them with confidence.

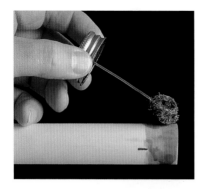

## 5 APPLY PRIMER.
Primer does two things for plastic pipe: It removes dirt and grease, and it softens the pipe surface slightly so the cement can form a better bond. Most primer has a built-in dauber. As you swab on the primer, be sure to cover the entire joint and fitting area.

## 6 APPLY CEMENT.
Make sure to buy cement for the specific material you're joining. Swab a generous amount around the pipe with the built-in dauber; then apply a thin coat of cement inside the fitting. Don't apply primer and cement to several pieces at one time—the cement sets up too fast.

## 7 JOIN AND TWIST.
Working quickly, insert the pipe in the fitting until it bottoms out. You'll get some excess cement oozing out near the joint. This is a good indicator that you've used enough cement. Adjust the pipe so your alignment marks are roughly one quarter-turn apart. Now twist the

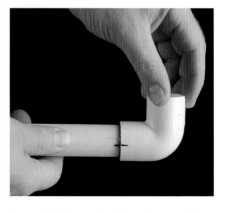

pipe the remaining quarter-turn until the marks align, to spread the cement inside the joint and ensure a good bond.

# Working with Flexible Pipe

**T**here are a number of flexible-tubing products that can make your adventures in plumbing easier. Flexible chrome pipe is used to connect exposed shut-off valves for hot and cold supply lines to faucets. Flexible copper tubing or "rolled" tubing is great for long runs, where sweating connections would be inconvenient. Flexible supply and waste lines, available in a variety of lengths and diameters, make connecting fixtures a snap; see pages 70–71.

**CUTTING FLEXIBLE TUBING.** Both chromed flexible supply lines and rolled copper tubing can be readily cut with a tubing cutter. Note: Since rolled tubing is thinner than rigid copper (that's why it's flexible), it doesn't take a lot of cutting-wheel pressure to cut through the pipe. Cut gently to keep from crimping the end of the tubing closed.

**BENDING FLEXIBLE TUBING.** To bend flexible tubing in a curve, especially a tight one, make sure to use a coil-spring tubing bender (available in the plumbing supply aisle). Slip the bender on the pipe and exert gentle hand pressure until the desired curve is achieved. If you try to bend this type of tubing without a bender, you'll likely kink the pipe, which will eventually fail and cause a leak.

# Flexible Supply and Waste Lines

Here they come to save the day—not Mighty Mouse, but flexible supply and waste lines! Although this sounds over-the-top, flexible supply and waste lines really can save the day. Quite often in a plumbing project, you just can't get from point A to point B with rigid pipe; odds are that you can with flexible piping.

**FLEXIBLE SUPPLY LINES.** Although chromed copper supply lines are somewhat flexible, braided metal and vinyl mesh supply lines like the ones shown in the photo

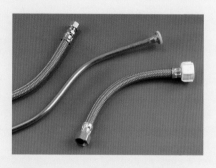

above right make fixture hookup a breeze. Flexible supply lines typically run between shut-off valves and fixtures—

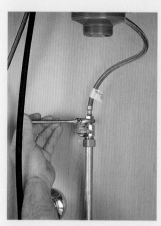

usually in tight, cramped spaces. Any flexibility in these situations is a plus. Flexible supply lines are available in a variety of pre-set lengths, complete with captive connecting nuts. When purchasing flexible supply lines, you're better off long than short; any excess can bend to one side. If they're way too long, you can even loop them.

**FLEXIBLE WASTE LINES.** Flexible waste lines connect sinks, tubs, and showers to the waste stack. Most flexible waste lines use plastic compression fittings to make their connections. You'll find a variety of flexible waste fittings in both 1¼" and 1½" diameters—they're easy to identify by their accordion-style mid-sections. These flex sections can be opened up or compressed (and angled) as needed. They're the quickest way to cure misalignment between a new sink and an existing waste line. Also, most flex waste fittings come with both 1¼" and 1½" compression washers so you can hook them up to either-diameter fittings, such as tailpieces and traps.

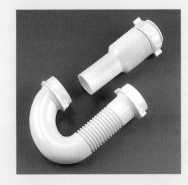

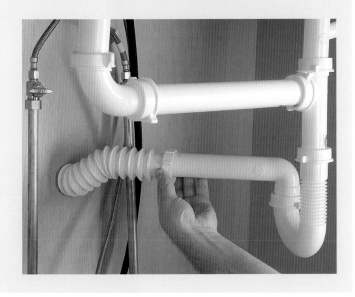

PLUMBING KNOW-HOW

# Working with Galvanized Pipe

The challenge to working with galvanized pipe is that the pipe is threaded on both ends. If you grip a pipe to turn it counterclockwise to loosen it from a fitting on one end, you're actually tightening the pipe into its fitting at the opposite end. Really. This means that to disassemble pipe, you have to start at the end of the run or a union (whichever is closer to where you're working), and work back to the section you want to replace. Ugh. No wonder copper pipe is so popular.

**USING A PIPE WRENCH.** To dismantle the pipe, position a pipe wrench on the pipe and another wrench on the fitting so their jaws are facing opposite directions. Move the wrenches toward the jaw opening to loosen; reassemble the system in the reverse order that you took it apart. Thread the parts together by hand and finish tightening them with pipe wrenches; be careful not to overtighten, or you'll crack the fittings.

**PIPE JOINT COMPOUND.** The last thing you want to happen after you've disassembled and then reassembled galvanized pipe is for your joints to spring a leak. Apply a generous dollop of pipe joint compound (page 80) to the threads of both the pipe and the fittings, as shown.

# Compression Fittings

**A** compression fitting is an easier alternative to a sweat fitting. Instead of solder, the watertight seal is created by the fitting itself. Although not as reliable as a soldered joint, a compression fitting is quick and easy — no heat, no mess.

**HOW COMPRESSION FITTINGS WORK.** A compression fitting consists of three parts: a fitting or body, a ferrule or compression ring, and a compression nut. The nut and ferrule slip over the pipe, which is inserted into the fitting. Tightening the nut compresses the ferrule into the fitting, creating a watertight joint.

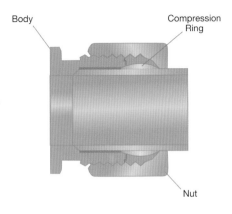

Body

Compression Ring

Nut

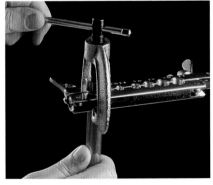

**FLARED FITTINGS.** A flare fitting is a special type of compression fitting that is commonly used for gas lines. The flared end of the pipe mates with a beveled edge inside the fitting. When a flare nut that slips over the pipe is tightened, a seal is formed. The flared end is made with a special flaring tool; see page 57.

# Rough-in Plumbing

If you'll be installing fixtures that require plumbing lines to be moved, you'll need to establish the placement of the new plumbing. The location and placement of piping is fairly uniform, but you should check the instructions for the individual fixtures. Most instructions provide rough-in dimensions—that is, where the piping needs to be located.

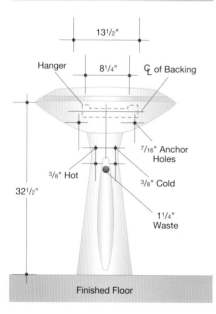

**VANITY SINK ROUGH-IN.** Typical supply and waste line locations for a sink mounted in a vanity are shown at left. Since the piping is concealed by the vanity, there's quite a bit of leeway here.

**WALL-MOUNT/PEDESTAL SINK ROUGH-IN.** Wall-mounted sinks are suspended from the wall (or a bracket attached to the wall). To support the weight of the sink, the framing behind the wall covering must be strengthened. This usually entails adding a support cleat between nearby wall studs. Pipe placement is critical here because in most cases the sink should conceal the pipes. What many folks don't know is that a pedestal sink is actually a wall-mounted sink. Yes, some of the weight is borne by the pedestal, but the main load is supported by the wall.

**TOILET ROUGH-IN.** Only two pipes are required for a toilet: the waste line and a cold-water supply line. The critical item here is waste-line placement. In years past when toilets were standard, the center for the closet flange was 12" from the wall (drawing at right). But newer toilets—particularly elongated bowl toilets (page 37)—require more clearance. Check rough-in dimensions before buying to make sure you have sufficient clearance.

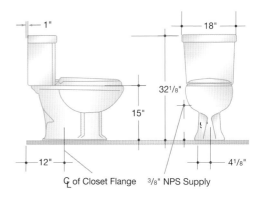

Ḡ of Closet Flange    3/8" NPS Supply

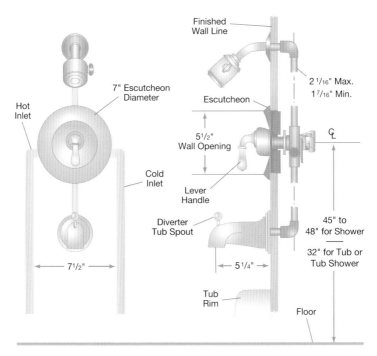

**TUB AND SHOWER ROUGH-IN.** For tubs and showers, the valve is located 45" to 48" above the floor for a shower, and 32" above the floor for a tub only. Showerheads are positioned about $6\frac{1}{2}$ feet above the floor as shown in the drawing above.

## Pipe access

Roughing in fixtures generally requires cutting access space for piping. There are two ways to do this: cutting notches or drilling holes. Which method you choose will depend on the diameter of the pipe and whether the pipe is traveling through vertical wall studs or a horizontal top or bottom plate. The key thing to remember is to keep the notch or hole as small as possible, to keep from weakening the framing. Also, whenever you run pipe through framing, you should always protect the piping by affixing a metal plate to the framing. This prevents accidental drilling or nailing into the pipe. You'll find these plates in the electrical aisle, since they're also used to protect wiring.

**CUTTING NOTCHES.** Hold the pipe against the stud and mark its position and width. Then make a pair of cuts to define the width of the notch, stopping at the desired depth; knock out the waste with a hammer, and trim with a chisel if necessary.

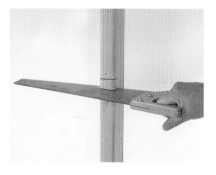

**DRILLING HOLES.** Fit the desired-diameter drill bit or hole saw in your drill, and drill the access hole centered on the width of the framing member.

# Securing Pipe

Whenever you run pipe inside a wall, you should always secure the pipe to the framing members. Securing pipe does two things: It keeps the piping from shifting over time (and causing problems), and it also helps support the weight of the piping and fixtures attached to it (such as a showerhead). The two most common ways to secure pipe are with pipe clips and with plumber's strap. In many cases, you'll need to add a cross cleat (see below) to fasten the clips or strap, as piping often runs in between studs.

**USING CLIPS.** Pipe clips or pipe straps are sized to match the diameter of the pipe that you're using. They're made of the same material as your pipe so there won't be any corrosion issues.

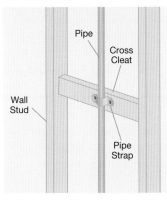

Pipe

Cross Cleat

Wall Stud

Pipe Strap

**USING PLUMBER'S STRAP.** Plumber's strap comes in rolls, since it's cut to length as needed. The strap has evenly spaced holes punched in it along its length to make it easy to fasten to framing members. Plumber's strap is commonly used to support pipe hanging from floor or ceiling joists.

**CROSS CLEATS.** To support piping between studs, cut a scrap of wood to span the studs and secure it to the studs with screws or nails. This provides a base for mounting pipe clips or plumber's strap.

# Creating Watertight Joints

The one basic skill essential for every homeowner plumber is creating a watertight joint. Fortunately, there are several easy-to-use sealants available that make this a straightforward task: Teflon tape, plumber's putty, pipe joint compound, and wax rings.

**TEFLON TAPE.** Ah...the miracle of Teflon. When DuPont created Teflon (then called polytetra-fluoroethylene)—the world's slipperiest material—in 1938, plumbing was likely the last thing on their minds. Used at first to create nonstick cookware, Teflon has over the years contributed to huge advancements in aerospace, communications, electronics, and architecture.

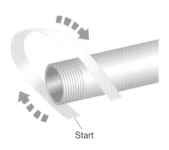

Start

Wrap Tape Clockwise So
Nut Threaded on End Will
Tighten Tape, Not Loosen It

Eventually it was discovered to be the ultimate sealant for threaded parts, and creating watertight joints became as simple as wrapping a few turns of Teflon tape around threads before screwing parts together.

There are a couple of rules for working with Teflon tape. First, direction is important. To prevent the tape from unraveling as the parts are threaded together, you should wrap the tape onto the threads as shown in the drawing above. Second, because this

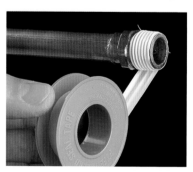

stuff is so thin, make sure to use enough—generally four to five turns will do the job. And finally, take care to wrap the tape evenly over the threads; don't let the tape bunch up (as it's prone to do), or you'll get an uneven seal.

**PLUMBER'S PUTTY.** Before the advent of another miracle—silicone—plumbers relied almost solely on plumber's putty to create watertight seals. Plumber's putty is still widely used today by pros who prefer its reliability and working properties over silicone. The problem they have with silicone is that its adhesive properties are too good. Silicone under a sink rim, for example, can make it diffi-

cult to remove the sink for repair or replacement. This is generally not the case with plumber's putty.

The dough-like putty is easily rolled into coils for wrapping around sink rims (photo below) and strainers. It's also excellent for preventing leaks under the base plate of a faucet. Many experienced plumbers will throw out the base plate gasket provided by the manufacturer and fill the cavity or gasket space in the underside of the base plate with putty, as shown in the photo above. Once the fixture is in place, excess putty is easily removed with a plastic putty knife.

**PIPE JOINT COMPOUND.** Before Teflon tape, plumbers relied on pipe joint compound or pipe "dope" to create watertight seals on threaded parts. The only hard and fast rule for applying this toothpaste-like compound is to use more rather than less. Too little compound results in leaks; excess compound is easily removed with a soft cloth. Although mainly used today for working only with galvanized pipe (the coarse threads on galvanized pipe tend to shred Teflon tape), pipe joint compound still has it uses.

Some pros apply a dollop of pipe dope to the threads of compression fittings, as shown in the photo below. Although it's technically not necessary, since the internal flange should create a

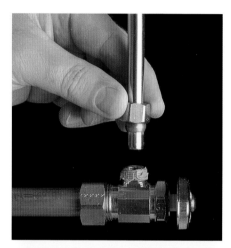

watertight seal, pros like the added anti-leak insurance that pipe joint compound adds to the fitting. Also, if you run out of Teflon tape and don't want to make a hardware store run, pipe joint compound can be used to join threaded copper parts, as shown in the bottom photo.

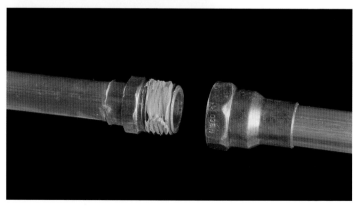

**WAX RINGS.** Wax rings are used to create a watertight seal between a toilet and its closet flange, as shown in the drawing below. There are two basic types of wax rings available: a simple ring of wax, and a ring with a rubber no-seep flange (photo at right). The no-seep rings offer added insurance against leaks.

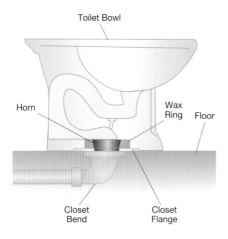

Toilet Bowl

Horn

Wax Ring

Floor

Closet Bend

Closet Flange

**PRO TIP**

## No-Wax Ring

An alternative to a wax ring is the "wax-free" ring (as shown here) manufactured by Fluidmaster (www.fluidmaster.com). Instead of wax and its mess, these rings rely on foam-rubber gaskets to create a seal.

# Removing a Sink

If it's time for a new sink, the old one, naturally, will have to go. Removing an old sink can be quick and easy or a real pain—it all depends on access to the sink and how the sink is mounted to the countertop. If the sink is mounted in a vanity, first remove the doors. This makes a restrictive space less restrictive and lets in light.

**1** **SHUT OFF WATER AND DRAIN LINES.** To remove a sink, first turn off the water supply to the faucet. If the sink has shut-off valves, turn the knobs clockwise. If not, shut off the hot

and cold water (the main valve and the water heater valve). Then open the faucet to drain out any water in the pipes.

**2** **DISCONNECT TRAP AND PLUG.** With a bucket under the trap, use slip-joint pliers to loosen the slip nuts that connect the trap to the tailpiece and waste line. Then carefully remove the trap and empty it into the bucket. Stuff a rag up into the waste line as shown to prevent sewer gas from seeping into your home.

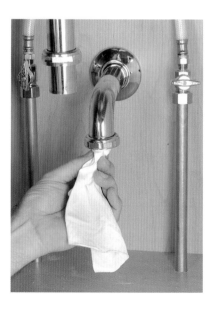

**4 LOOSEN HARDWARE OR SEAL.** If the sink you're removing is truly self-rimming (as shown here), run a putty knife under the rim to break the seal. For sinks mounted with clips under the countertop, remove these first before freeing the rim. Note that old corroded clips may have to be cut or snapped off to free them from the sink.

**3 DISCONNECT SUPPLY LINES.** Next, disconnect the faucet supply lines running to the shut-off valves, using an adjustable wrench, and let them drain into the bucket, as shown.

**5 LIFT OUT THE SINK.** For heavy sinks, enlist the aid of a helper to lift it out of the vanity or countertop. Have the helper push up from below so you can grip the sink's rim. If the sink is cast iron, have scraps of wood on hand to slip under the rim to keep from crushing your fingers. With the sink out, you'll have complete access to the faucet—see pages 84–85 on how to remove it.

# Removing a Faucet

Removing a faucet can be easy or hard—it all depends on access. Although it may sound like a lot of extra work, the way to get complete access to the faucet is to remove the sink it's mounted to, or remove the vanity top it's attached to. If you don't do either of these, you'll end up on your back, reaching up behind the sink to loosen the mounting nuts. At the same time you have to navigate past the supply and waste lines, working in an area that provides little, if any, elbow room. Fortunately, there's a nifty tool that can alleviate some of the problem; see the sidebar below. Before you can remove your old faucet, you'll need to turn off the water and disconnect the supply lines leading to the faucet (see page 83).

## QUICK FIX

## Using a Basin Wrench

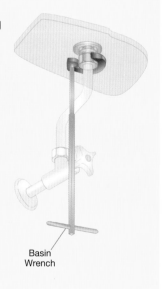

A basin wrench is a specialty tool that's designed to let you reach up and loosen faucet nuts in close quarters. It has a long extension arm that lets you loosen or tighten faucet-mounting nuts in the clear space below a sink. With access so limited, it's hard to adjust the wrench to fit the nut. That's why the serrated jaws of a basin wrench are self-adjusting—they close to fit around the nut as pressure is applied. You'll find basin wrenches wherever plumbing supplies are sold.

Basin Wrench

**1** **LOOSEN FAUCET-MOUNTING NUTS.** Faucets mount to sinks and countertops in a variety of ways. Centerset faucets typically use large nuts that thread onto the posts in the base. Other faucets (like the one shown here) attach via a mounting bracket held in place with a nut. Whatever the mounting system, you'll have to remove it in order to pull out the faucet. For restricted spaces, use a basin wrench; see page 84.

**2** **BREAK SEAL UNDER BASE PLATE.** Before you remove the faucet, run the blade of a putty knife around its perimeter to sever the bond between the faucet and sink from the old caulk or plumber's putty. Note: If you're replacing the faucet on a bathroom sink, you'll need to disconnect the pop-up mechanism that controls the drain stopper before you can remove the faucet. (See pages 130–133 for more on pop-up mechanisms.)

**3** **LIFT OUT FAUCET.** Grasp the faucet firmly and pull up. This may take some muscle—the putty or caulk used to install the original faucet often develops a surprisingly strong bond over time.

**4** **CLEAN BENEATH FAUCET.** Once you've pulled out the old faucet, it's important to remove any old putty or caulk from the sink. If you don't, you may not get a good seal under the new faucet. Use a putty knife to scrape away the bulk of the old sealant. Then clean the surface thoroughly with a soft rag.

# Removing a Toilet

Although removing a toilet isn't one of the daintiest parts of a plumbing job, it's pretty simple. Just make sure to have plenty of towels and rags on hand to clean up water spills, and a helper to remove and dispose of the old toilet. (Even when separated into two parts, toilets are surprisingly heavy.)

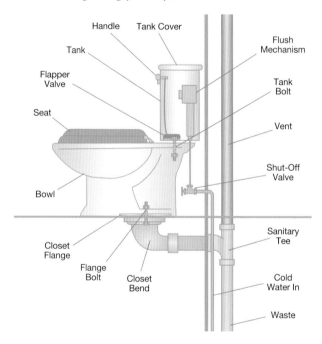

**TOILET ANATOMY.** There are two main parts to a standard toilet: the tank and the bowl. The tank holds water to flush the existing contents of the bowl down the waste line (1.6 gallons in all toilets made after 1996, and up to 3 gallons in toilets made prior to 1996). When the flush handle is depressed, the lever arm raises a ball or flapper in the bottom of the tank, allowing the water in the tank to flow down into the bowl. The water flows through a series of holes in the rim and with the aid of a little gravity, forces the contents of the bowl to exit through the integral trap, out through the horn, and down into the waste line via a sanitary tee.

**1 SHUT OFF THE WATER.**
To remove a toilet, begin by shutting off water to the toilet and emptying the tank completely. Flush the toilet and leave the handle depressed to empty as much water out of the tank as possible. Then sop up the remaining water with a sponge. Next, use an adjustable wrench to loosen the supply line to the toilet. Then unhook this line from the shut-off valve.

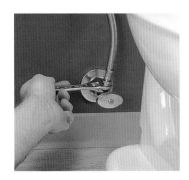

**2 LOOSEN THE TANK BOLTS.** Although it's not absolutely necessary, it's always a good idea to remove the tank from the bowl to lessen the weight of the toilet. Just the weight of the bowl alone is enough to strain most backs. The tank is held in place by a set of bolts inside at the bottom. These bolts pass through rubber washers and into a ledge of the bowl, where they're held in place with nuts. Hold each bolt securely and loosen the nut with a wrench.

**3 REMOVE THE TANK.**
Once you've removed both tank bolts, lift the tank straight up off the bowl as shown and set it aside.

**4** **LOOSEN THE CLOSET FLANGE BOLTS.** Next, pry off the decorative caps at the base of the toilet that cover the closet bolts and mounting nuts. The nuts that thread onto the closet bolts can be difficult to get off. Try loosening them with an adjustable wrench, as shown. If they don't come off easily, apply some penetrating oil and allow it to soak 15 minutes before trying again. If that doesn't do the trick, you'll have to cut the nuts off with a hacksaw, as described in the sidebar below.

Q  U  I  C  K    F  I  X

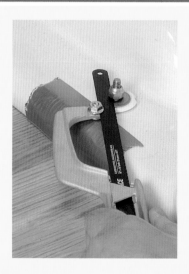

# Stubborn Closet Flange Bolts

The mounting nuts that thread onto closet bolts have a well-deserved reputation for not coming off easily. If penetrating oil doesn't work, you'll have to cut the nuts off with a hacksaw. The best tool for this is actually a mini-hacksaw, as shown, since the end of the blade can be flexed to cut in tight quarters.

**5** **LIFT OFF BOWL AND SET ASIDE.** Now lift the toilet gently off the floor. Since there's sure to be water still remaining in the integral trap, be careful as you move it about; empty this water into a bucket, and set the toilet on an old towel. If you find that the toilet won't come up easily, try rocking it gently from side to side to break the old seal.

**6** **REMOVE OLD WAX RING.** Plug the drain (see the sidebar below) and remove the old closet flange bolts from the closet flange; set them aside if you're planning to reuse them. Then use a putty knife to scrape away the old wax from both the closet flange and the underside of the toilet around the horn. Wipe any remaining wax away with a clean, soft rag.

## PLUG THE CLOSET BEND

➕ As soon as you've set the toilet aside, plug the closet bend opening with a rag to prevent sewer gas from escaping into the house. Make sure the rag is large enough that it can't accidentally fall down into the waste line.

# Installing a Shut-Off Valve

**E**very fixture in your home should have its own shut-off valve. If this isn't the case, it's worth spending the time to install them. When an emergency occurs, or when it's simply time to make a repair, you (and your family) will appreciate the fact that you can do the job without having to turn off the water for all, or even part, of the house.

**1** **CUT THE PIPE.** To install a shut-off valve, turn off the supply to the house (including the hot water line), drain the lines, and then cut the pipe with a tubing cutter (page 62) where the valve is to be installed.

**2** **SWEAT ON A FITTING.** Sweat-type fittings require more work than compression fittings but will provide a better seal. Start by sweating either a male or female transition fitting onto the pipe to match the valve.

# 3 APPLY TEFLON TAPE.

When the fittings are cool to the touch, wrap a couple turns of Teflon tape around the threads, as shown. At the same time, wrap a couple turns of Teflon tape around the threaded end of the valve that will accept the flexible supply lines added later. It's a lot easier to do this now, before the valve is in place.

# 4 SCREW ON THE SHUT-OFF VALVE.

Now you can install the shut-off valve by threading it onto the fitting. Whenever possible, adjust the valve's position so that the control handle is easily accessible.

# 5 CONNECT THE SUPPLY LINES.

Finally, connect the newly installed shut-off valves to the faucet (or other fixture), using flexible supply lines as shown.

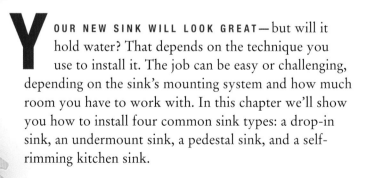

# 5

# Installing Sinks

**Y**OUR NEW SINK WILL LOOK GREAT—but will it hold water? That depends on the technique you use to install it. The job can be easy or challenging, depending on the sink's mounting system and how much room you have to work with. In this chapter we'll show you how to install four common sink types: a drop-in sink, an undermount sink, a pedestal sink, and a self-rimming kitchen sink.

# Connecting Supply Lines

Every time you install a new sink, you'll need to hook up supply lines to the faucet and the waste line to the trap. In this chapter we show you how to install four different types of sinks—each mounts to a countertop in a unique way. What's not unique to each sink is the supply and waste hookup—it's pretty much the same for all. To avoid repetition, we'll cover how to connect supply lines on pages 94–95, and how to connect waste lines on pages 96–99.

As a general rule, it's easiest to hook up the supply lines first, since the waste line will block access to the supply lines. Typically, you'll connect the hot and cold shut-off valves to the hot and cold valves of the faucet, as shown in the drawing below. These connections are usually made with flexible supply lines—either chromed copper or braided or plastic flexible lines with captive nuts (see page 70).

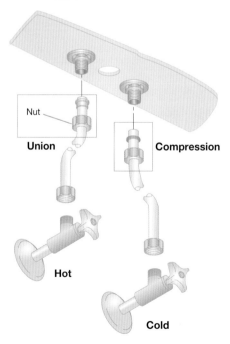

Nut

**Union**

**Compression**

**Hot**

**Cold**

**1** APPLY TEFLON TAPE TO FAUCET VALVES. Since it's difficult to install flexible supply lines to a faucet once the sink is in place, it's best to attach these to the faucet valves when you have full access to the faucet. Start by wrapping a few turns of Teflon tape around the threads of the hot and cold valves.

**2** ATTACH FLEXIBLE SUPPLY LINES TO FAUCET. Now you can thread the faucet end on each of the flexible supply lines onto the hot and cold valves and tighten them with an adjustable wrench.

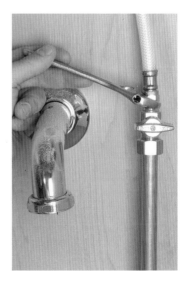

**3** CONNECT TO SHUT-OFF VALVES. Next, wrap the threads of the shut-off valve with Teflon tape, and thread on and tighten the nuts with an adjustable wrench.

# Connecting Waste Lines

How a waste line connects to a sink depends a lot on the type of sink. Sinks with single drains are much simpler to plumb than sinks with two or more drains, like many kitchen sinks. As a general rule, chrome-plated brass fittings hold up better over time than PVC plastic fittings that are connected via compression fittings, and they're less susceptible to puncture. The exception to this is ABS plastic that's cemented together; see page 99 for more on this.

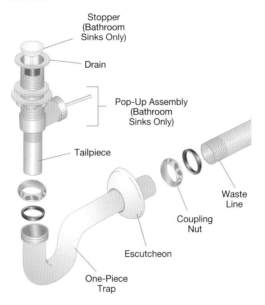

Stopper
(Bathroom
Sinks Only)

Drain

Pop-Up Assembly
(Bathroom
Sinks Only)

Tailpiece

Waste
Line

Coupling
Nut

Escutcheon

One-Piece
Trap

**TYPICAL WASTE HOOKUP.** Sinks with single drains connect to the waste line via a tailpiece and a trap, as shown in the drawing above. Bathroom lavatories commonly have a pop-up mechanism (see pages 130–133) that can open and close the drain as needed. Kitchen sinks with multiple drains are more complicated and usually involve running a tailpiece down from each strainer. These tailpieces are then connected together with either one or two elbows and a wye (Y) connector.

**1** **PLUMB OUT FROM WALL.** Depending on your existing waste line setup, you may or may not need to run a new trap line to connect to the trap. In many cases, you'll be connecting the new waste line to an existing line that has been cut off; misalignment here is a common problem. Two solutions: flexible waste line parts (see page 71), and flexible rubber transition fittings. Flexible transition fittings have a hose clamp on each end; one is tightened around the new line, and the other is tightened around the existing waste line.

**2** **ADD EXTENSIONS (IF NECESSARY).** Additionally, you may need to extend the tailpiece(s) down far enough to reach the existing trap line. Extensions in various lengths are available wherever plumbing supplies are sold, and are easily cut to length with a hacksaw.

**3** **CONNECT THE TRAP.** With the extension(s) in place, you should be able to connect the trap to the extension and trap line, as shown. In general, you'll want to push the trap up onto the extension until the lower end mates with the trap line.

**4** **TIGHTEN THE FITTINGS.** For the trap line, trap, and tailpiece, start by tightening the slip nuts by hand. Then grip each coupling nut with a pair of sliding-jaw pliers and give each nut an additional quarter-turn.

## FLEXIBLE LINES ON DISPOSERS

Although flexible waste lines may help cure misaligned parts, there are situations where it's best not to use them. A prime example: When there's a garbage disposer in use. The accordion-style body that lets you flex and angle the part also tends to catch ground-up food particles, resulting in frequent clogs. If you do have a kitchen sink with a garbage disposer and an alignment problem, take the time to re-plumb the waste line, using non-flexible waste line parts.

# Pro Waste Lines

Pros rarely, if ever, use the white plastic traps and waste line parts found in most home centers and hardware stores. There are two reasons for this. First, the walls of these products are quite thin and prone to puncture. Second, the parts are commonly joined together with plastic compression fittings that tend to fail over time—especially on kitchen sinks equipped with a garbage disposer. Even the highest-quality garbage disposer will vibrate in use, and the vibration will eventually loosen the compression fittings and cause the joints to fail. So what do pros use? Heavy-duty ABS parts that cement together—the only threaded fittings are for the trap to allow you to remove it for cleaning.

**1 CEMENT TRAP SECTIONS.** Pros plumb out from the wall with ABS parts that are cemented together. These parts have much thicker walls and won't puncture. And because the parts are cemented together, leaks aren't an issue. This type of waste line connection can even stand up well to disposer use.

**2 ATTACH THREADED TRAP.** The trap can't be cemented to the waste line because it wouldn't allow access for cleaning or augering. That's why it connects to the other parts via threads.

INSTALLING SINKS

# Installing a Drop-In Sink

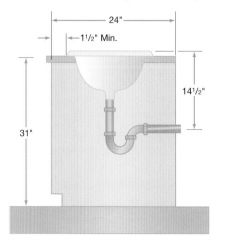

24"

1¹/₂" Min.

14¹/₂"

31"

**A** drop-in sink is a bathroom classic. To achieve a watertight seal, a bead of silicone is run around the rim before placing the sink. In most cases, the weight of the sink and the silicone are what hold it in place. But be careful here: The adhesive properties of silicone will make it tough to remove the sink later without damaging the vanity top. Plumber's putty also forms a good seal and will let you easily remove the sink at a later date.

Installing drop-in sinks is easier than in the past because they've been standardized to fit in pre-cut holes in ready-made vanity tops. For a new vanity top-and-sink combination, check to make sure they fit together before installation. In some cases you'll be installing a sink in a new vanity top where you'll have to cut an opening. If you're installing the sink in an existing vanity top and you need to enlarge the opening, see page 101.

**1 LAY OUT THE OPENING.** Many sink manufacturers provide a paper pattern of the hole you'll need to cut out in the vanity top. If one is supplied with your sink, cut it out and tape it to the vanity top so it's centered from side to side, and the desired distance from the front. Next, trace around the pattern and then remove it.

**2 CUT OUT THE OPENING.**
With the cutout marked,
drill a starter hole for a saber
saw blade. Then use a saber
saw fitted with a wood-cutting
blade to cut the hole. Note that
we applied a layer of masking
tape to the bottom of our saw

to prevent its metal base from scratching the laminate countertop.
Take your time and cut to the inside (or waste portion) of the line.

**P R O   T I P**

# Enlarging an Existing Opening

If you're replacing a sink, but not the countertop, the
new sink may be too large for the existing opening.
If you have a router or laminate trimmer, you can accurately
enlarge the opening easily. The challenge to enlarging an
opening just slightly is it's tough to take just a bit off accurately. Start by running a $1/4$" rabbeting bit around the inside
of the opening. Then follow this up with a patternmaker's
bit. The bearing of the patternmaker's bit will track along the
rabbet you just cut to create an accurate, larger opening.

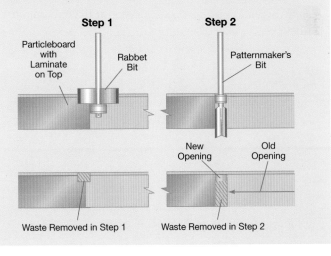

Step 1 — Particleboard with Laminate on Top — Rabbet Bit

Step 2 — Patternmaker's Bit — New Opening — Old Opening

Waste Removed in Step 1

Waste Removed in Step 2

INSTALLING SINKS

**3** INSTALL THE FAUCET. Now is the best time to install the faucet because you have full access to the underside of the sink. For instructions for a centerset faucet, see pages 126–129; widespread faucets are covered on pages 134–137.

**4** APPLY A SEALANT. If the drop-in sink you're installing doesn't rely on mounting clips, apply a generous bead of silicone around the edge of the opening, as shown here. For other sinks, apply a coil of plumber's putty.

**P R O  T I P**

# Preventing Countertop Damage

■▲ Most countertops covered with plastic laminate are made from particleboard. Although particleboard is great for this since it's very flat and bonds well to laminate, unprotected particleboard soaks up water like a sponge. That's why it's a good idea to seal the edges of the opening with a couple coats of glue or latex paint. This is added insurance that if the seal under the sink fails, the countertop won't swell.

**5** **INSTALL THE SINK.** With the faucet in place, lift the sink and set it into the opening. Press down to compress the sealant. Adjust the sink from side to side and front to back.

**6** **WIPE OFF ANY EXCESS SEALANT.** Once you've set the sink, wipe up any excess sealant immediately. Use a dry, clean cloth or one you've moistened with the recommended solvent. Allow the sink to sit undisturbed overnight to let the silicone set up. The next day, hook up the supply and waste lines; see pages 94–95 and 96–99, respectively. When complete, remove the aerator and flush the system.

# Installing an Undermount Sink

**W**ith no rim, an undermount sink presents just a smooth, clean sweep of countertop. There's no entry point for water to leak through, and no countertop crevices for dirt to hide in. There is, though, a limitation: Undermount sinks can be installed only where the countertop has a *solid* edge—as with solid-surface materials, like Corian, or solid wood. You can't use a laminate countertop here, because the exposed plywood or particleboard edges will soak up moisture and swell.

Whether your sink mounts under a wood top (as shown here) or a solid-surface material, you'll need to cut an opening in the vanity top. In either case, the edges of the opening will be seen and must be cut perfectly smooth and true. With wood, this can be done with a template and router. But since most solid-surface materials can be cut only by certified fabricators, you'll need to send them either the sink or a pattern before you order the vanity top, so they can custom-cut the opening.

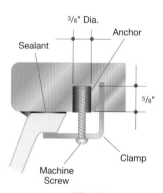

**Marble**

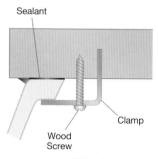

**Wood**

**MOUNTING DETAILS.** The clamps that hold undermount sinks to the underside of the countertop are secured in one of two ways, as shown in the drawing at left. On solid-surface and stone countertops, shallow holes are drilled and fitted with threaded inserts or anchors. With a solid-wood countertop, the screws can be driven directly into the wood; no inserts are required.

**1** **MARK THE OPENING.**
Undermount sinks come with a paper pattern of the opening you'll need to cut. In order to rout the finished edge later, it's best to make a template of the pattern from $1/4$" hardboard. Carefully cut out the pattern and tape it to the hardboard. Then trace

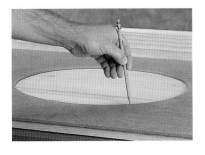

around it, remove the pattern, and cut out the opening, taking care to stay on the waste side of the cut. Then go back and sand to the line; a drum sander works great for this. Place the completed template on the vanity top so it's centered from side to side and the desired distance from the front, and trace around the opening.

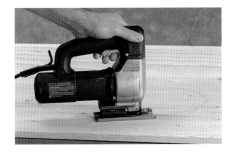

**2** **CUT OUT MAJORITY OF WASTE.** Drill an access hole for a saber saw and cut out the majority of the opening waste, staying on the waste side of the marked line.

**3** **ROUT THE FINISHED EDGE.** Now place the template back onto the top and clamp it in place so it's aligned with the line you marked earlier. Fit a laminate trimmer or router with a pattern-maker's bit (same as a flush-trim bit, except the bearing is on top of the cutter instead of the bottom), and rout the finished edge. Alternatively, you can sand to the marked line with a drum sander.

**4** INSTALL THE MOUNTING HARDWARE. Most undermount sink makers detail the exact position of the mounting clamps. We marked our hardboard template and used this to transfer the clamping clip locations onto the underside of the top. Note that we used a simple depth gauge (masking tape) attached to the bit to keep from drilling through the top.

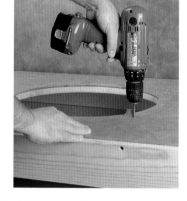

**5** APPLY A SEALANT. Although the sink rim is actually under the countertop, you still want a watertight seal between the sink and the top to prevent leaks. Apply the recommended sealant to the perimeter of the opening as directed.

**6** SET AND SECURE SINK. Now you can carefully set the sink in place so it's centered on the opening. Your best bet here is to place the vanity top on a pair of sawhorses so you can look up from underneath to make sure the sink is in the desired position. When in place, secure the sink with the rim clamps provided.

**7** **DRILL HOLES FOR THE FAUCET.** Since you can't attach a faucet to an undermount sink, you'll need to drill a hole or holes in the countertop as detailed in the faucet installation instructions.

**8** **INSTALL THE FAUCET.** All that's left is to add the faucet and install the top on the vanity. For a centerset faucet, see pages 126–129. Step-by-step installation instructions for a widespread faucet are on pages 134–137. With the faucet and pop-up assembly installed, flip the top upright and position it on the vanity. Adjust it from side to side and from front to back, and insert shims as needed to level it. Then secure the top to the vanity with the screws provided. For wood tops, make sure to seal the top and especially the rim edges with multiple coats of marine-grade polyurethane or spar varnish.

# Installing a Pedestal Sink

**W**ith its slender stem supporting a basin, a pedestal sink looks elegant to most people. What's more, a pedestal sink can make a small space bigger, especially in a bathroom that previously housed a vanity. On the downside, when you install a pedestal sink, you'll generally lose storage space. And, a pedestal sink is more complicated to install than it looks: You need to remove the wall covering to install a cleat to support the sink, as shown in the drawing on page 109. That's right—a pedestal sink is basically a wall-mount sink. All the pedestal does is partially obscure the waste line and trap—and look good.

Contrary to the "beauty" shots in plumbing catalogs, most pedestal sinks do not eliminate having to look at plumbing lines. (Those catalog shots are photographed in a studio; the fixtures are never hooked up.) The supply lines on a pedestal sink that's actually installed are connected to shut-off valves typically located on both exterior sides of the pedestal—they're often highly visible, as shown in the drawing below.

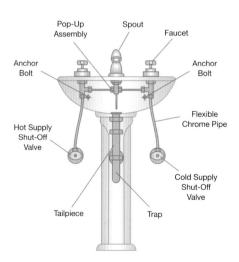

**FRONT VIEW OF A PEDESTAL SINK**

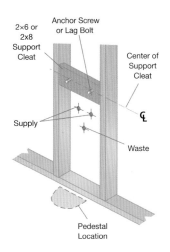

2×6 or 2x8 Support Cleat

Anchor Screw or Lag Bolt

Center of Support Cleat

Supply

℄

Waste

Pedestal Location

## PEDESTAL FRAMING

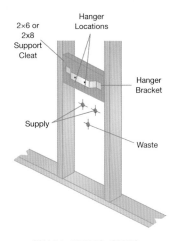

2×6 or 2x8 Support Cleat

Hanger Locations

Hanger Bracket

Supply

Waste

## WALL-HUNG-SINK FRAMING

**1 NOTCH THE STUDS.** If your old sink wasn't wall-mounted, you'll need to remove the wall covering, install a cleat, and replace the wall covering, as illustrated in the drawing above. Consult 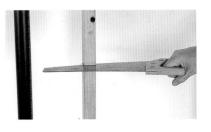 the installation directions that came with your sink to locate the cleat. Measure up the recommended distance and cut partially through the wall studs, as shown. Then knock out the waste with a hammer and clean up the notches with a chisel as needed.

**2 CUT AND INSTALL THE BRACE.** Cut a brace (typically a 2×6 or 2×8) to fit between the studs. Attach the brace securely to the wall studs with nails or, better yet, screws. Note: In some cases, you'll need to re-plumb the supply and waste lines before replacing the wall covering.

## 3 RE-APPLY THE WALL COVERING.

With the brace in place, you can reinstall the wall covering. Measure carefully and cut a piece of drywall to fit. Then measure and lay out holes for the supply and waste lines, and cut these with a drywall saw or hole saw. Position the drywall on the wall, and secure to the studs with drywall screws or nails.

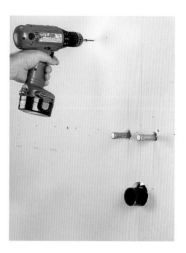

## 4 MARK THE PEDESTAL LOCATION.

Locate the centerline of the sink. Then use a framing square as shown to transfer the centerline onto the floor where the pedestal will stand.

## 5 POSITION THE PEDESTAL.

Now carefully set the pedestal on the floor the recommended distance from the wall so it's centered on the centerline you just marked on the floor.

**6** **ADD THE SINK.** Now you can lift the sink into position and slowly lower it onto the pedestal, as shown. Adjust the sink as needed from side to side so it rests firmly on the pedestal. Adjust the pedestal location as needed—just be sure to keep it centered on the centerline.

**7** **LEVEL THE SINK AS NEEDED.** Next, check to make sure the sink is level. If necessary, shim the pedestal or sink. Since shimming the pedestal will be noticeable, most pedestal sink manufacturers suggest applying self-adhesive rubber gasket material between the top of the pedestal and the underside of the sink to level the sink.

INSTALLING SINKS

**8** **MARK MOUNTING HOLES.** With the sink leveled and positioned properly, you can mark the mounting hole locations. The sink will attach to the wall and the support cleat in one of three ways: It'll hang on a bracket attached to the cleat, it'll be fastened to the cleat with hanger bolts (as shown here), or it'll use both. Whichever method is used, mark the location of the bracket, or mark through the mounting holes of the sink onto the wall, as shown. On many sinks, the pedestal is secured to the floor though a hole in its base—mark this hole location now.

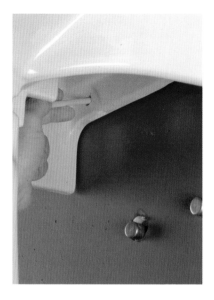

**9** **DRILL THE SINK MOUNTING HOLES.** Next, drill the recommended-sized holes for the screws or bolts that attach the sink to the wall, as shown. Make sure that you hit the brace that you installed earlier.

**10** **INSTALL THE HANGER BOLTS OR BRACKET.** Now you can install the hanger bolts or bracket that will support the sink. Brackets typically bolt directly to the wall (and to the brace behind the wall covering). The simplest way to install hanger bolts is to drive them in with a socket wrench, as shown and as described below. Note that most manufacturers will define the maximum distance the bolt can protrude from the wall (in our case, it's 1 inch).

## Installing Hanger Bolts

◤◥ A hanger bolt has coarse threads on one end that ◣◢ are driven into framing, and fine threads on the other end to accept a nut. So how do you install them? Just thread two nuts onto the finely threaded end, as shown. Use a pair of wrenches (or a wrench and a socket) to tighten the nuts firmly against each other. This will let you drive the bolt into the framing. Once in place, use the wrenches to free the nuts and unthread them from the hanger bolt.

**12** **INSTALL THE SHUT-OFF VALVES.** Now's also the time to install shut-off valves. For more on this, see pages 90–91.

**11** **DRILL THE PEDESTAL MOUNTING HOLE.** Before you can install the pedestal and sink, there are a few more things to do. First, if your pedestal is designed to attach to the floor, drill the recommended-sized hole in the flooring for the bolt or screw that secures it in place.

**13** **INSTALL THE FAUCET.** It's also easiest to install the faucet before setting the sink in place. For a widespread faucet, see pages 134–137; directions for installing a centerset faucet begin on page 126.

**14 INSTALL THE DRAIN.** Finally, you'll want to install the pop-up mechanism now and adjust it for proper operation before setting the sink. For more on pop-up mechanisms, see pages 130–133.

**15 INSTALL THE PEDESTAL.** With all the prep work done, you can install the pedestal and sink. Start by setting the pedestal in place. Check to make sure that it doesn't interfere with the supply or waste lines, or with the mounting bracket if installed.

INSTALLING SINKS

**16** **SET THE SINK IN PLACE.** Now you can position the sink on the pedestal. As you do this, take care to slip the mounting holes in the back of the sink over the hanger bolts you installed earlier.

**17** **SECURE THE SINK AND PEDESTAL.** Check one more time that the sink is level. Then slip washers and nuts over the hanger bolts and tighten them hand-tight. Then go back and give them a quarter-turn with an adjustable or socket wrench. Don't tighten any more than this or you might crack the sink. Thread the pedestal mounting bolt or screw through the mounting hole in the pedestal, and secure the pedestal to the floor.

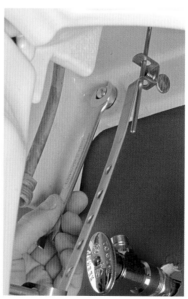

**18** **CONNECT THE SUPPLY LINES.** Now you can connect the supply lines. If these will be visible, consider using chrome supply tubes to connect the faucet valves to the shut-off valves. Measure and cut these as needed with a tubing cutter. Thread the flanged end through the faucet mounting nuts and secure these to the faucet; make sure to wrap a couple of turns of Teflon tape around the valve threads. Then insert the cut end of the tubes into the shut-off valves and tighten the compression fittings. Alternatively, connect these with flexible supply lines, as shown here.

**19** **CONNECT THE WASTE LINE.**

Connecting the waste line and trap to a pedestal sink is one of the great challenges in plumbing. This is because you're working in such tight quarters. Whenever possible, hook the trap up to the tailpiece before setting the sink in place. This way all you'll have to do is connect the trap to the trap line—and this is usually an exposed connection that you can get to, as shown here.

# Installing a Kitchen Sink

The self-rimming stainless steel sink shown here is a kitchen classic. The term "self-rimming" simply means that the lip or flange of the bowl overhangs the countertop. This in itself doesn't create a watertight seal. To achieve that, the sink has clips that hook onto a lip on the underside of the sink to pull the sink down tight against the countertop. This presses the bead of sealant or putty under the rim firmly against the countertop to prevent leaks.

**1 TRACE AROUND THE SINK.** If you're installing a sink in a new countertop, you'll have to first cut an opening. Although many sink manufac-turers provide paper templates, some don't. The sink we used here didn't come with a pattern, so we had to use the sink itself to lay out the opening. Start by positioning the sink on the coun-tertop and then trace around it. You'll need to mark the opening smaller than this so the sink's rim rests on the countertop; see the sidebar below.

**P R O  T I P**

## Reducing Sink Patterns

To create the smaller pattern you'll need for the sink opening, set a compass to the amount of rim overhang (typically ½" to 1"). Hold the non-marking tip of the compass on the sink outline. As you move the compass around the outline, the marking tip will scribe the perfect sink opening.

**2** **CUT OUT THE OPENING.** With the opening marked, drill a starter hole for a saber saw blade inside the opening near the line you marked earlier. Then use a saber saw fitted with a wood-cutting blade to cut the hole. Note that we applied a layer of masking tape to the

bottom of our saw to prevent its metal base from scratching the laminate countertop. Take your time and cut to the inside (or waste portion) of the line.

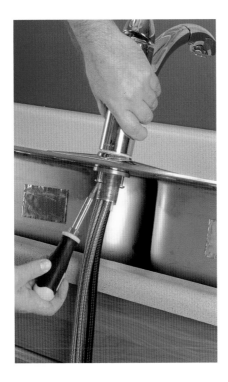

**3** **INSTALL THE FAUCET.** It's easiest to install the sink faucet before installing the sink because you have better access. (See pages 138–141 for more on installing a kitchen faucet.) Set the sink upside down on the countertop, and slide it so it overhangs the countertop enough for you to insert the faucet from underneath. Use the gasket supplied or plumber's putty to create a seal. Secure the faucet by tightening the mounting nuts.

**4** **INSTALL THE MOUNTING CLIPS.** There are about as many types of sink mounting clips as there are manufacturers of self-rimming sinks. Virtually every sink maker has a unique clip and mounting system. Regardless of the style, all clips for self-rimming sinks attach to a flange around the sink's perimeter. And most manufacturers are very specific as to where these clips should be installed. An example of this is the drawing below. Note the slightly offset locations on the sink shown.

With some sinks, you can snap the clips onto the flange (as shown here) before installing the sink in the opening. Other sinks require you to hook on the clips once the sink is in place.

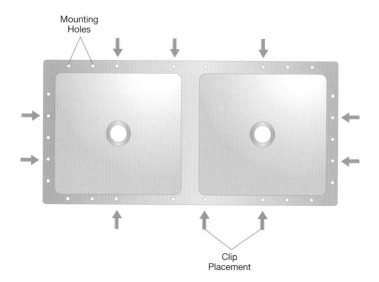

Mounting
Holes

Clip
Placement

**5** **ADD THE SEALANT.** Apply a sealant under the rim of the new sink. Use either a continuous, generous bead of silicone caulk or a $\frac{1}{2}$"-diameter coil of plumber's putty around the rim. Alternatively, you can apply the sealant to the edge of the sink opening (as shown here). In either case, you should make sure the existing countertop is free of old putty or caulk by first wiping it down with mineral spirits or acetone.

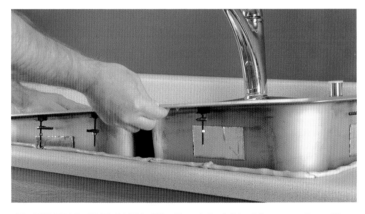

**6** **INSTALL THE SINK.** Flip the sink right-side up and position it in the opening. Center it from front to back and from side to side before pressing down to compress the sealant. If you're installing a cast iron sink, it's a good idea to temporarily place scraps of wood on each side of the opening and set the sink on these. Cast iron sinks are heavy, and these wood scraps help prevent squished fingers.

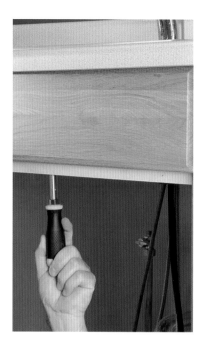

## 7 TIGHTEN THE MOUNTING CLIPS.

With the sink in place, install the rim clips per the manufacturer's directions (if they are not already installed), and tighten them to lock the sink in place. As a general rule, you'll want to work alternate edges of the sink, tightening a clip on one side, then the clip opposite it. This helps ensure an even squeeze-out of the sealant. Remove any putty or caulk squeeze-out from around the rim with a plastic putty knife; then wipe off the sink and countertop with a clean, soft cloth.

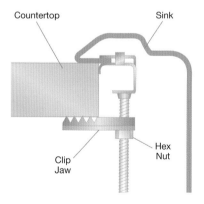

Countertop

Sink

Clip Jaw

Hex Nut

**8** **INSTALL THE STRAINERS.** Most strainers have five parts: the strainer body, a rubber gasket, a fiber gasket, a strainer nut, and a strainer basket. Though you might think the rubber gasket will create a watertight seal, it won't. You'll need to apply plumber's putty around the drain opening's lip (top photo). Then pass the strainer body through the sink and press it firmly into the putty. On the underside of the sink, slip on the rubber

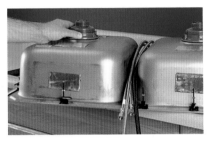

and fiber gaskets and the strainer nut. Hand-tighten the nut and fully tighten with a strainer wrench; see the sidebar below. Some find it easier to install the strainers before installing the sink (middle photo). Finally, hook up the supply and waste lines; see pages 94–99).

Finally, hook up the supply and waste lines; see pages 94–99).

**P R O  T I P**

## Strainer Wrenches

A strainer wrench has two different notched ends to fit over the ribs in either a three- or four-ribbed strainer body. To use one, hold the strainer nut with one hand. Use the other hand to turn the strainer wrench to pull the strainer body tight against the drain opening. Remove any excess putty with a plastic putty knife.

INSTALLING SINKS

# 6

# Installing
# Faucets

**T**HE FAUCET LEAKS, or looks dated, or doesn't have the features that you want. For any or all of these reasons, replacing a faucet is one of the most common plumbing projects for do-it-yourselfers. Faucet installation is straightforward, but it can be tricky since you're usually working in tight quarters. Fortunately, there are ways to get around this. In this chapter we'll show you how to install centerset, widespread, and kitchen faucets.

# Installing a Centerset Faucet

Although a simple job, installing a new faucet can be challenging. The tricky part—like changing the oil filter in many cars—is accessing the parts. Because of the location of the faucet, you'll probably end up on your back reaching up behind the sink to loosen the faucet mounting nuts. At the same time you have to navigate past the supply and waste lines, working in an area that provides little, if any, elbow room. That's why we recommend that you install a faucet with the sink out of the vanity or countertop. In some cases, it's easier to pull off a vanity top than it is to replace a faucet with the sink in place.

Since a centerset faucet is basically one piece (drawing at left), it's the simplest faucet you can install. The valves on most centerset faucets are spaced on 4" centers—something to keep in mind when shopping for both a sink and a faucet. Many newer sinks have 8"-on-center spacing and require the use of a widespread faucet (see page 134).

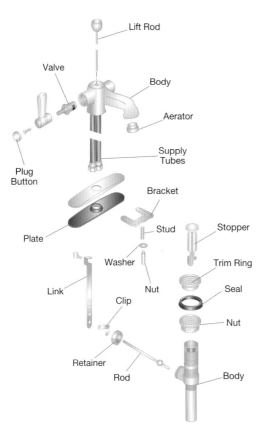

Lift Rod

Valve

Body

Aerator

Supply Tubes

Plug Button

Plate

Bracket

Stud

Stopper

Washer

Trim Ring

Link

Nut

Seal

Clip

Nut

Retainer

Rod

Body

**1** CLEAN THE SINK. If you're replacing a faucet, take the time once you've pulled out an old faucet (pages 84–85) to remove any old putty or caulk from under the faucet. If you don't, you may not get a good seal under the  new faucet. Use a plastic putty knife to scrape away most of the old sealant. Then clean the surface thoroughly with a soft rag and some denatured alcohol.

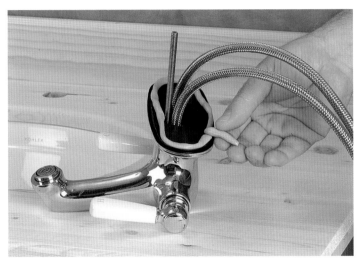

**2** PREPARE THE FAUCET. To install a centerset faucet, start by preparing the faucet. To create a watertight seal between the base plate of the faucet and the sink, slip the gasket provided onto the bottom of the faucet. Alternatively, use a trick that many plumbers use—throw away the gasket and pack the cavities under the base plate with plumber's putty. This will take quite a bit of putty, but it's inexpensive and will create a lasting seal.

INSTALLING FAUCETS

**3 INSERT THE FAUCET.**
Once you've applied plumber's putty or a gasket, slip the faucet through the hole (or holes) in the sink.

**4 SECURE THE FAUCET.**
Then thread the mounting nuts (or bracket nut, as shown here) onto the faucet. Check to make sure that the faucet is centered on the sink, and fully tighten the valve nuts or bracket nut. If you packed the faucet cavity with plumber's putty, remove any squeeze-out with a plastic putty knife.

**5 INSTALL THE DRAIN BODY.** Next you can install the drain body in the sink. Start by applying a coil of plumber's putty around the lip of the drain or under the lip of the drain flange. Push the drain body up against the underside of the sink drain and screw the drain flange onto the drain body, as shown. Remove any excess putty.

**6** **HOOK UP THE POP-UP MECHANISM.** With the drain body in place, install the rest of the pop-up mechanism parts and adjust the mechanism for proper operation. See pages 130–133 for detailed directions on installing a pop-up mechanism. If you've installed the faucet with the sink removed, now install the sink. Then connect the faucet to the shut-off valves using chromed supply lines or flexible supply lines; connect the tailpiece to the trap and existing waste line (see pages 94–99).

**PRO TIP**

## Remove Aerator Before Testing

Whenever you install a new faucet, particularly if you've altered the supply lines, you should always remove the faucet's aerator before turning on the water. That's because there is frequently debris in the lines that would quickly clog up the aerator. With the aerator removed, turn on the water and flush the system. When the water runs clear, turn off the faucet and reinstall the aerator.

# Installing a Pop-Up Mechanism

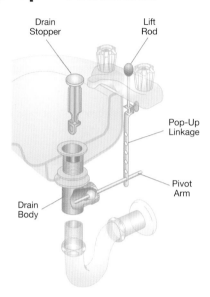

Drain Stopper

Lift Rod

Pop-Up Linkage

Pivot Arm

Drain Body

**A** bathroom lavatory uses a pop-up mechanism to open and close the sink's drain. A pop-up mechanism has four main parts: the drain body, the lift rod, the linkage, and the drain stopper, as shown in the drawing at right. The lift arm is connected to the pop-up linkage, which controls the pivot arm. The drain stopper hooks up to the pivot arm. Raising or lowering the handle of the lift arm forces the pivot arm to move up and down, allowing the stopper to open or close the drain.

**1 SEAL THE DRAIN.** On most pop-up mechanisms, the drain flange screws onto threads cut into the top of the drain body. The seal between the drain flange and the sink is formed with plumber's putty. Apply a generous coil under the lip of the flange, as shown.

## 2 INSTALL THE DRAIN BODY.
Insert the drain flange through the drain opening in the sink. Thread this into the drain body and then tighten the nut on the drain body to push its rubber gasket up against the bottom of the drain opening. Tighten this nut with a pair of slip-joint pliers to create a watertight seal. Remove any excess plumber's putty around the drain flange.

## 3 INSTALL THE PIVOT ROD.
Insert the stopper through the drain flange into the drain body. Align the slot in the end of the plunger with the opening in the drain body for the pivot rod. Then insert the washer and pivot rod into the opening in the drain body so the end of the pivot rod passes through the slot in the end of the plunger. Slip the plastic nut over the open end of the pivot rod and thread this into the drain body. Tighten to friction-tight, and check the plunger's action by pivoting the rod up and down—the plunger should open and close as you do this.

## 4 INSTALL THE TAIL-PIECE.
Wrap a few turns of Teflon tape around the tailpiece and screw it into the drain body.

INSTALLING FAUCETS

**5** INSTALL THE LIFT ROD. Next, push the lift rod through the faucet's base plate. Connect the rod to the pop-up linkage by passing its end through the U-shaped bracket on one end of the pop-up linkage. Tighten the thumbscrew friction-tight to lock the pop-up rod in place.

**6** ATTACH THE LINK TO THE ROD. Slip one of the holes in the linkage over the end of the pivot rod and secure it with the spring clip. Bend the extension rod as necessary so one of its holes aligns with the pivot rod provided. The spring clip is designed to capture the pop-up linkage between its ends, as shown.

**7** ADJUST AS NEEDED.
Check the plunger
action by moving the lift
rod up and down. Adjust
the extension as necessary
for smooth operation. If
the drain stopper doesn't
fully close the drain, the
pivot rod may need to be
adjusted. Start by pinching
the spring clip that con-
nects the pivot rod to the
lift arm. Pull the lift arm off
the pivot rod and insert the
rod in a different hole in
the lift arm. A little trial-and-
error testing will be needed
here to determine which
hole in the lift arm works
best. If, after you've
adjusted the pivot rod,
the stopper still doesn't
operate properly, adjust
the lift rod. Loosen the
thumbscrew on the
linkage and push the
linkage up farther on
the link rod. Tighten
and test; readjust
as necessary.

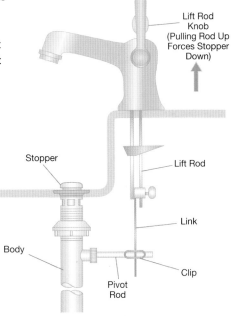

Lift Rod
Knob
(Pulling Rod Up
Forces Stopper
Down)

Stopper

Lift Rod

Link

Body

Clip

Pivot
Rod

# Installing a Widespread Faucet

Unlike a centerset faucet (pages 126–129), where the handles and spout all share a common base, a widespread faucet is made up of separate components (drawing below.) Separate components allow more flexibility in placing the spout and handles. (Note: These faucets can also be installed in any sink designed for a centerset faucet.) Widespread faucets are more complicated to install than one-piece units, as the individual components need to be connected with tubing. On some faucets, a T-connector hooked up to the spout accepts hot and cold water from valves installed beneath the faucet handles. Other systems run hot and cold water directly into the handle, which then routes the mixture off to the spout via a flexible line.

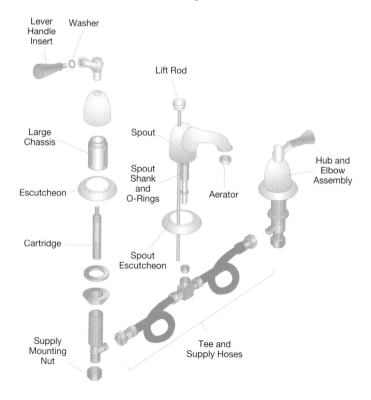

Lever Handle Insert

Washer

Lift Rod

Large Chassis

Spout

Spout Shank and O-Rings

Hub and Elbow Assembly

Escutcheon

Aerator

Cartridge

Spout Escutcheon

Supply Mounting Nut

Tee and Supply Hoses

**1 APPLY PLUMBER'S PUTTY UNDER THE SPOUT.** The first part to install on a widespread faucet is the spout. You can use the gasket provided to create a seal or pack the underside of the spout with plumber's putty, as shown here.

**2 ATTACH THE SPOUT.** Slip the spout through the center hole in the sink. The spout is held in place with a flange and nut. Hold the spout firmly in position, thread on the washer and nut supplied with the faucet, and tighten the nut with an adjustable wrench. Some manufacturers supply a socket that can be used with a screwdriver to tighten the nut; others provide a wrench just for this task.

INSTALLING FAUCETS

**3** INSTALL THE VALVES. If the widespread faucet you're installing uses faucet valves to route the water to the spout, they can be attached next. Some types are inserted up through the sink; others go in from above. Use the mounting nuts supplied by the manufacturer to secure the faucet valves. Adjust the orientation of the valves as shown in the installation directions.

Faucet manufacturers generally indicate which way the handles of the valves should point, as shown in the drawing on page 137. Once in position, tighten the mounting nuts firmly with a large adjustable wrench or slip-joint pliers.

PRO TIP

## Identifying Valves

Every widespread faucet has two unique valves, one hot and one cold. Although they may look the same, the internal parts are different: The hot valve has special parts designed to handle hot water. So it's crucial that you install the correct valve in its correct position: hot valve to the left of the spout, cold valve to the right. Faucet manufacturers want to make sure you install the valves correctly. Some will label the cold valve (as shown here); others use the universal colors of red for hot and blue for cold on the valve supply connectors.

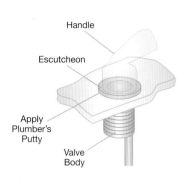

Handle

Escutcheon

Apply
Plumber's
Putty

Valve
Body

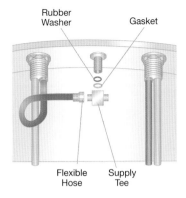

Rubber
Washer

Gasket

Flexible
Hose

Supply
Tee

**4 ATTACH THE T-CONNECTOR TO THE SPOUT.** If a T-connector is used to connect the spout to the faucet valves, wrap a few turns of Teflon tape around the threads of the spout and thread on the T-connector. Use an adjustable wrench to tighten it in place, as shown.

**5 CONNECT THE VALVES TO THE SPOUT.** Now you can connect flexible lines between the T-connector and the hot and cold valves—odds are you'll need to coil each into a loop to make the connection. Make sure to wrap a couple of turns of Teflon tape around the threads before attaching these; then tighten the nuts with an adjustable wrench. On some faucets, the flexible lines are permanently connected to the T; others must be attached to both the T-connector and then to the valves. Finally, hook the supply lines to the faucet, as described on pages 94–95.

**Installing a Widespread Faucet** 137

# Installing a Kitchen Faucet

**P**ull-out sprayer faucets combine a sprayer and a faucet in one, so they free up holes in the sink or countertop for other accessories like a soap dispenser or instant hot-water dispenser. They also have the added benefit of providing spray action at the touch of a button. Pull-out sprayer faucets are installed much the same way as a standard faucet, but with a few peculiarities.

A sprayer faucet consists of a base plate, a faucet body with flexible supply tubes, the pull-up sprayer, a hose that connects the sprayer to the faucet body, and a counter-weight that helps retract the sprayer hose when it has been pulled out of the faucet body; see the drawing at left.

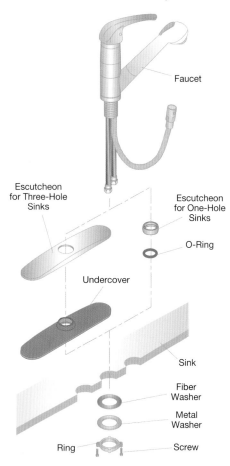

Faucet

Escutcheon for Three-Hole Sinks

Escutcheon for One-Hole Sinks

O-Ring

Undercover

Sink

Fiber Washer

Metal Washer

Ring

Screw

**1** **PREPARE THE BASE.** Use the gasket provided with the new faucet, or discard the gasket and pack the base with plumber's putty, as shown here. Plumber's putty does a better job of sealing base plates of faucets against uneven surfaces—something that your average gasket can't handle.

**2** **INSERT THE FAUCET IN THE SINK.** On most pull-out sprayer faucets, the faucet body is first installed in the sink and secured, and then the sprayer hose is threaded through the body and attached to the base. On other faucets this isn't necessary, and all you need do is thread the unit through the opening in the sink or countertop.

**3 SECURE THE FAUCET.** Mounting systems for kitchen faucets vary greatly from one manufacturer to another. Typically, some type of bracket or large washer threads onto the base of the faucet body. The brackets usually house a pair of screws or bolts that, when tightened, pull the faucet body firmly into its base plate and the sink top (as shown here).

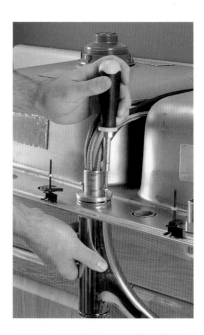

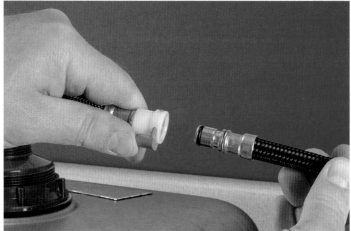

**4 CONNECT THE SPRAYER.** Once the faucet is secured, connect the sprayer hose to the faucet body. On many faucets, you'll thread a captive nut on the end of the sprayer onto the designated flexible tubing coming out of the faucet body for the sprayer. Other faucets use a nifty quick-connect fitting, as shown here.

**5** **ATTACH THE COUNTERWEIGHT.** Most pull-out sprayers come with a counterweight that you attach to the sprayer hose to retract the hose back into the faucet body once it's been extended. The counterweight is usually a pair of lead weights that are split to wrap around the hose and are screwed together. Take care to tighten them only friction-tight, as you could restrict water flow by overtightening.

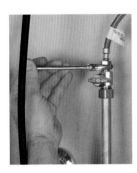

**6** **CONNECT THE SUPPLY LINES.** Whenever possible, attach flexible supply lines to the faucet first before mounting the sink. Wrap fresh Teflon tape around the threads of the shut-off valves of the water supply lines, and tighten the connecting nuts with an adjustable wrench.

**P R O   T I P**

## Preventing Sprayer Hose Interference

The counterweight that retracts a sprayer hose is notorious for catching on anything under the sink. One way you can help reduce the likelihood of this is to reduce potential obstacles. One example: Pull the flexible supply lines out of the way (as shown here), and hold them in place with an electrical-cable tie.

# Plumbing a Shower Faucet

The plumbing for a stand-alone shower is simpler than that for a bathtub/shower, naturally: There's no line running down to a bathtub spout. There's just a hot and cold supply line running to the valve and a pipe running up to the shower arm and showerhead (drawing below). The valve may use one or two handles to adjust the temperature. The metal escutcheons that cover shower valves tend to be large so that you have full access to the plumbing lines in case repairs are needed. Newer valves use smaller escutcheons and assume either that you have an access panel behind the shower or that you'll remove the wall covering if repairs are necessary.

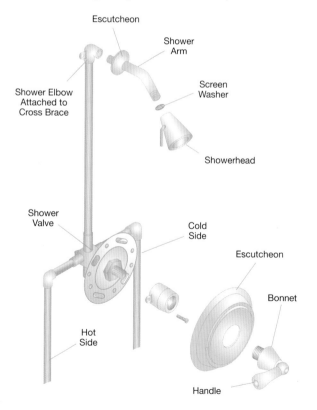

Escutcheon

Shower Arm

Shower Elbow Attached to Cross Brace

Screen Washer

Showerhead

Shower Valve

Cold Side

Escutcheon

Bonnet

Hot Side

Handle

**1** **REMOVE THE OLD VALVE.** To install a new shower valve, turn off the main water shut-off and drain the lines. Then use a tubing cutter or hacksaw to cut the existing hot and cold lines running to the valve. Free the showerhead fitting from the cleat or framing that it's attached to, and remove the old valve.

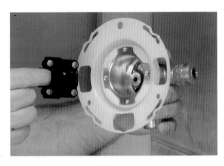

**2** **PREPARE THE NEW VALVE.** If you're planning on sweating pipe directly to the new valve, you should always remove the inner workings of the valve since they'd be damaged by the heat of the torch. See the valve installation manual for valve preparation instructions. On the other hand, if you're using thread-on couplings (as we did here), removing the inner valve parts won't be necessary.

**3** **ADD A SUPPORT CLEAT.** Odds are that you'll need to install a support cleat to span the wall studs and create a mounting platform for attaching the new plumbing. The thickness of this cleat will depend on the finished position of the valve; see your valve installation manual for specifics.

**4** **PLUMB THE NEW VALVE.** Plumbing the new valve may be as simple as adding a couple of copper couplers and a few lengths of copper. Or it may be as complex as redoing both lines to meet the new valve location. For detailed instructions on working with copper pipe, see pages 62–65. Note: Make sure to use a heat shield (as shown here) to protect existing framing and wall coverings (see page 49).

**5** **INSTALL THE VALVE.** If you removed the inner valve parts to sweat the valve onto new copper lines, go ahead and replace the inner valve once the valve body has cooled. Alternatively, if you're using thread-on couplings (as shown here), connect the valve to the threaded fittings using slip-joint pliers to fully tighten the connections.

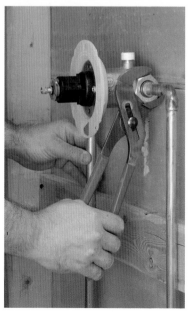

**6** **SECURE THE VALVE.**
At this point you can
secure the valve to the support cleat or framing. How
you do this will depend on
the location of the valve
and the cleat. Some valves
have mounting holes in the
body for attaching the valve
directly to a cleat. Valves
with mounting holes (like
the one shown here) can be
secured by attaching the
supply lines to the cleat with
pipe straps.

**7** **PLUMB THE SHOWER**
**LINE.** With the valve in
place, you can plumb the
shower line. In most cases
this is simply a length of
copper pipe with a male
adapter on one end and a
shower elbow on the other.
The length of the pipe will
depend on where you want
the showerhead. To mini-
mize cutouts in shower wall
enclosures, many plumbers
position the shower elbow
above the shower wall
(as shown here). Once
you've sweated the parts
together, wrap a few turns
of Teflon tape around the
male adapter and use an
adjustable wrench to tighten
it into the valve.

**8** STUB OUT THE SHOWER LINE. Since you'll probably be covering the walls with drywall (see below) and will need to conceal joints with joint compound over a period of a couple of days, thread a stub (basically a threaded pipe with a cap on one end) into the shower elbow. This lets you restore water to the house and not have to worry about accidentally turning on the shower as you drywall and mud. This is a good time to add the plaster guard for the valve; see the sidebar on page 147.

**9** INSTALL THE WALL COVERING. With all the plumbing complete, you can install the wall coverings. Moisture-resistant drywall is installed first, followed by your waterproof covering of choice. This can be a solid-surface surround, a fiberglass surround (as shown here), or tile. (For more on installing a shower unit, see pages 160–169.)

# 10 REMOVE THE SHOWER LINE

**STUB.** Once all the wall coverings are in place and complete, you can remove the shower line stub in preparation for installing the showerhead.

## Plaster Guards

Pros always use the plaster guards that come with valves to protect the valves when wall coverings are installed. A plaster guard is basically a plastic dome (or other shape) that snaps over the valve, as shown here. Keep the plaster guard in place until you're ready to complete the installation by adjusting water temperature limits and adding the escutcheon and shower handle(s).

INSTALLING FAUCETS

## 11 INSTALL THE SHOWERHEAD.

Slip the escutcheon over the shower arm and wrap a few turns of Teflon tape around both threaded ends of the arm. To install the showerhead, start by screwing the arm into the shower elbow; tighten and adjust its position so it points directly down. Then screw the head onto the arm and tighten it with an adjustable wrench. Finally, push the escutcheon up against the wall, as shown.

## 12 ADJUST THE WATER TEMPERATURE LIMIT. Most

newer shower valves have some sort of built-in temperature limit to prevent scalding. The limit is similar to a governor on
a motor—it restricts how far you can turn on the hot water. See
your installation manual for specifics.

**13** INSTALL THE HANDLE ADAPTER (IF NECESSARY).
Many tub and shower valves require an adapter between the valve stem and the shower knob or handle. If your valve requires an adapter (as ours did), install it now.

**14** INSTALL THE ESCUTCHEON AND HANDLE. Finally, you can install the trim and handle(s). Start by attaching the escutcheon. For the valve shown here, the escutcheon is secured to the valve adapter with screws. Next, install the handle. The handle shown here just slips over the handle adapter and is locked in place with a setscrew. Restore water and check for proper operation.

# 7

# Installing Fixtures

**U**NLIKE, SAY, A COAT OF WALL PAINT, fixtures are intended to be fairly permanent. No one we know regularly replaces toilets or showers just for fun, so when you do update or upgrade, you want your work to last. That's why installing fixtures like a toilet, shower, shower door, and water dispenser takes special skills and knowledge. (We've already covered sinks, a separate kind of fixture, in Chapter 5.) In these pages, we'll take you through what you need to know to get those new fixtures in correctly for long-lasting results.

# Installing a Toilet

**A**lthough not one of the more glamorous plumbing projects, installing a new toilet is something that everyone in the family will appreciate—especially if it has any of the newer features: raised-height seat, elongated front, or pressure-assist flush mechanism.

There are two main parts to a toilet: the tank and the bowl (drawing below). The tank holds a preset amount of water to flush the existing contents of the bowl down the waste line (1.6 gallons in all toilets made

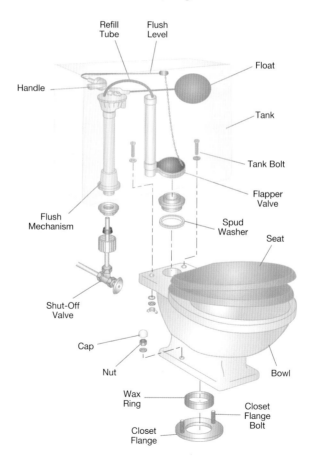

Refill Tube

Flush Level

Float

Handle

Tank

Tank Bolt

Flapper Valve

Flush Mechanism

Spud Washer

Seat

Shut-Off Valve

Cap

Nut

Bowl

Wax Ring

Closet Flange Bolt

Closet Flange

after 1996, and up to 3 gallons in toilets made before 1996). When the flush handle is depressed, the lever arm raises a ball or flapper in the bottom of the tank (by way of either a chain or lift wires), letting the water in the tank flow down into the bowl. The water flows through a series of holes in the rim and, with the aid of a little gravity, forces the contents of the bowl to exit through the integral trap. Then the waste moves out through the horn, and down into the waste line via a sanitary tee.

**1 CHECK THE FIT.** The first thing to do when installing a new toilet is to unpack the bowl and check the fit. Although most bowls will fit on a standard closet flange installed 12" away from the wall, some won't. So before proceeding, place the bowl over the closet flange to make sure that it'll fit.

**2 INSTALL THE WAX RING.** Position the wax ring so that it's centered on the closet flange, and press it down onto the flange. The wax ring shown here is the no-seep variety.

## 3 INSTALL THE CLOSET BOLTS.

Next, insert closet flange bolts (sometimes referred to as "Johnny" bolts) into the slots in the closet flange. The T-shaped head is inserted into the larger opening on the flange and then slid around the curved slot to the correct position. Bolts should be positioned on opposite sides of the flange. These bolts pass up through holes in the bowl's base.

## 4 POSITION THE BOWL.

With the seal in place, remove the rag in the closet flange used to keep out sewer gas (if it hasn't been removed previously; see page 89), and position the bowl over the closet bolts. Have a helper assist you in gently lowering the bowl in place so the closet flange bolts pass up through the mounting holes in the base of the bowl.

**5** SECURE THE BOWL TO THE CLOSET FLANGE. Thread the mounting nuts on the bolts and alternately tighten each nut until the bowl is flush with the floor. A word of caution here: Overtightening these bolts can crack the toilet—if they feel snug, leave them be.

## Setting a Bowl

**◤◥** Position the bowl over the closet bolts and gently lower the bowl; a helper is invaluable in getting everything aligned. Once in place, press firmly down on the bowl—don't stand or jump on it, as this will only overcompress the wax ring, resulting in a poor seal.

INSTALLING FIXTURES

**6** **PLACE THE TANK ON THE BOWL.** With the bowl in place, the next step is to attach the tank. Flip the tank upside down on the bowl and check to make sure the spud washer is in place. If it's not, install it now. Then turn the tank over and set it on the bowl so the spud washer is centered on the inlet opening. Align the holes in the tank with the holes in the bowl, and insert the tank bolts.

**7** **SECURE THE TANK TO THE BOWL.** To tighten the tank bolts, insert the tip of a long screwdriver in the slot in the bolt inside the tank. Thread on the nut by hand until it's snug. Then switch over to a socket wrench or adjustable wrench to finish tightening. Proceed with caution, since overtightening can crack the tank.

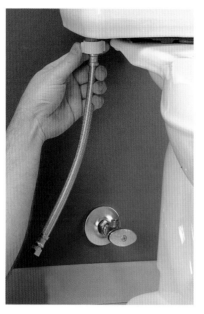

**8** CONNECT THE SUPPLY LINE TO THE TOILET. On most toilets, the nut that secures the top of the supply line to the tank is plastic— tighten these by hand only, since pliers can crack them quite easily.

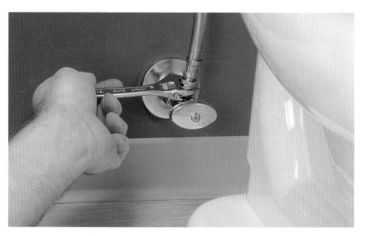

**9** CONNECT THE SUPPLY LINE TO THE SHUT-OFF. Now you can hook the tank up to the water supply. You can use flexible line for this (as shown here), or install a chrome supply tube if the supply line is highly visible. Just make sure to wrap a few turns of Teflon tape around the threads of the shut-off valve before threading on and tightening the nut.

**10** ADJUST THE FILL LEVEL (IF NECESSARY). Now you can turn on the water and test for leaks and proper flushing action. Check to make sure the tank is filling up to the recommended level (usually cast into the inside face of the tank). If your toilet uses a float cup, you can raise or lower the cup (and the water level, respectively) by first pinching the metal retaining clip that fits over a pull rod. Then slide the clip (and cup) up or down to the desired water level and release the clip. Flush the toilet and check the level. Readjust as necessary.

**11** ADD THE TANK COVER. When the fill level looks good and the toilet is flushing properly, add the cover to the top of the tank.

**12** **INSTALL THE SEAT.** The seat can now be attached to the bowl. It's held in place commonly with a pair of screws or bolts and nuts. Many seats also come with a pair of double-sided adhesive pads that fit under the plastic hinges of the seat to help prevent the seat from shifting from side to side in use. Once you've threaded a nut onto a screw or bolt, hold the nut secure and tighten the screw or bolt with a screwdriver.

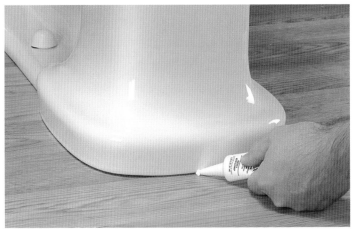

**13** **SEAL AROUND THE BASE OF THE BOWL.** All that's left is to caulk around the base of the toilet. Not only is this added protection against leaks, but it also prevents water from seeping under the toilet when you mop the floor. A high-grade silicone caulk is best for this, as it will remain flexible over time, keeping the seal intact.

# Installing a Shower

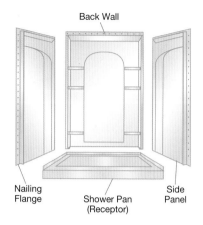

Back Wall

Nailing Flange

Shower Pan (Receptor)

Side Panel

**F**ixture manufacturers have made installing retrofit, standalone showers easy by making the shower unit modular. (The average one-piece shower unit will be too bulky to fit through most exterior doors, hallways, and bathroom doors.) Unlike its one-piece new-construction cousin, a modular shower typically consists of four pieces, as shown in the top drawing: a shower pan (often called a receptor, in the trade), a back wall, and two side panels. The most challenging part of the job is setting the pan or receptor so it's fully supported and level. The back wall and side panels usually snap into each other and the receptor, and are secured to the framing; see below.

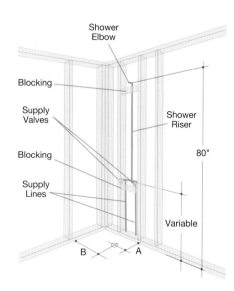

Shower Elbow

Blocking

Supply Valves

Blocking

Supply Lines

Shower Riser

80"

Variable

B      A

**TYPICAL SHOWER FRAMING.** Stand-alone showers need a frame to support the walls of the shower, whether they're made of acrylic panels or tiles. Consult the fixture installation instructions for the recommended size and layout of the framing. Besides the frame that supports the shower walls, you'll need to add cleats to support both the shower valve and the showerhead; see pages 142–149.

# 1 INSTALL THE DRAIN IN THE RECEPTOR.

The first step to installing a stand-alone shower is to install the drain in the hole in the receptor. In most cases, this means threading one half of a two-part drain assembly through the hole in the bottom of the shower pan and into the other half that is attached to the waste line later.

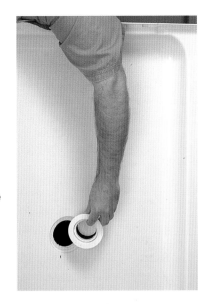

# 2 A MORTAR FOUNDATION.

To fully support and level a receptor, most shower manufacturers suggest setting the receptor in a bed of mortar. If you're planning on this, now is the time to mix the mortar and apply it to the floor. If you choose not to use mortar to support the receptor, you'll need to support it with shims or with what many plumbers use now: non-expanding foam (see page 162).

**3** **COVER THE MORTAR.** Once the mortar has been applied, cover it with a thin plastic drop cloth. The drop cloth prevents the mortar from sticking to the underside of the receptor. This makes it easy to remove in the future, and lets you remove it now (if necessary) to add more mortar without making a mess.

**4** **ADJUST THE RECEPTOR FOR LEVEL.** Carefully lower the receptor over the bed of mortar and check to make sure it's level. Press down as needed to redistribute the mortar. If necessary, remove it and apply more mortar. This will be the foundation for your shower, so take your time here and make sure it's fully supported and level.

P R O  T I P

## Bed of Foam

Instead of mortar, some pros support receptors (and tubs) by filling gaps between the receptor or tub and the subfloor with *non*-expanding foam. Note: Do not use standard expanding foam for this—this type of foam will expand and raise the tub or receptor completely off the floor so it's not level anymore.

**5** SECURE THE RECEPTOR TO THE FRAMING. Allow the mortar to set up overnight before securing the receptor to the framing. How you secure the receptor will depend on the unit. You'll either drill holes in the flange at the stud locations for screws or nails, or you'll use large-headed roofing nails or screws to secure the flange at its edge (as shown here).

**6** LOCATE THE PLUMBING OPENINGS. With the plumbing in place (pages 142–149), you can transfer the faucet locations onto the side panel and drill the appropriate-sized holes. Be especially exact in transferring these locations, as there is little, if any, margin for error. Use a hole saw (page 164) or saber saw to cut these openings. If you use a saber saw, drill access holes first and be sure to apply masking tape to the underside of the saw's base to prevent it from scratching the side panel.

**7** PRE-FIT THE WALLS. Once you've cut the plumbing openings, check to make sure the side panel fits over the plumbing. Then insert the back wall and other side panel to make sure everything fits before you move on to actual installation. A helper is extremely useful here, since you'll likely need to lift the unit up and over the receptor and navigate around any plumbing lines.

**PRO TIP**

## Smooth Plumbing Openings

◤◣ When pros need to drill holes greater than 1½" in diameter, they reach for a hole saw. Not only does a hole saw cut a perfect circle, but it also leaves relatively smooth edges. The most common type of hole saw has two parts: a twist bit that guides the cut, and a metal cup with saw teeth that do the cutting. The cup is usually carbon steel or bimetal. Regardless of the type you use, make sure to use slow speeds and stop often to blow out dust that can clog up the teeth and cause burning.

**8 CAULK THE RECEPTOR.** When you're sure the back and side panels fit perfectly, remove them and thoroughly clean the ledge of the receptor where the back wall and side panels rest. Additionally, many manufacturers will have you caulk all or part of the receptor before positioning the back wall and side panels; see your installation manual for specifics.

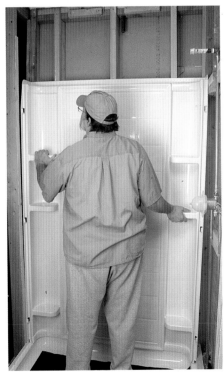

**9 INSTALL THE BACK WALL.** Now you can lift the back wall in place and set it onto the receptor. On many shower units, there are tabs molded into the bottom edge of the back wall and side panels that fit in corresponding notches in the receptor. These both lock the wall or panel in place and align it.

**10** INSTALL THE SIDE PANELS. With the back wall in place, lift and place the side panels in position. These pieces not only have tabs in the bottom to align them with the receptor, but they also have tabs on their long edges that fit into notches in the back wall to lock the pieces together.

**11** SHIM THE PANELS (IF NECESSARY). Before you can secure the back wall and side panels to the framing, check to make sure that there's no less than $1/8$" gap between the top edge of the wall or panels and the framing. If you find gaps greater than $1/8$", use pairs of shims to fill the gaps.

**12** SECURE THE PANELS TO THE FRAMING. Now you can secure the back wall and side panels to the framing, using the recommended fasteners and technique described in your installation manual. For the unit shown here, we used large-headed screws.

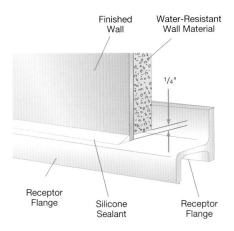

Finished Wall · Water-Resistant Wall Material

1/4"

Receptor Flange · Silicone Sealant · Receptor Flange

**Shower Receptor with Drywall**

# 13 INSTALL WALL COVERING.

With the shower unit in place, the next step is to cover the exposed studs with moisture-resistant drywall. But there's a problem here—the flange of the back wall and side panels. The drywall has to be installed over this and will go in at an angle. To prevent this, most pro installers will attach shims to the wall studs that are the same thickness as the flange. These can be cut from scraps of 2-by material and go up quickly. With the shims in place, cut and install the drywall and finish as desired. Note that most manufacturers will provide details such as the recommended gap between the finished wall covering and the receptor, as shown in the drawing above.

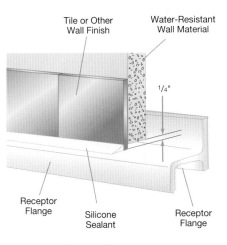

Tile or Other Wall Finish · Water-Resistant Wall Material

1/4"

Receptor Flange · Silicone Sealant · Receptor Flange

**Shower Receptor with Tile, Marble, etc.**

**14** **CAULK AS DIRECTED.** Once the wall covering is in place, apply caulk as directed by the manufacturer to create a seal between the back wall and side panels and wall covering.

**15** **CONNECT THE DRAIN.** Now is also the time to connect the drain in the receptor to your waste line. For homes with access under the shower, this is simply a matter of connecting the drain to the trap via a tailpiece. On a home built on a concrete slab (as shown here), there in no access below. So, you'll need to use a com- pression-type connector that fits over the existing waste line: When tightened, the connector compresses over the pipe to create a watertight seal.

## 16 ADD THE DRAIN COVER.

Most drains have a pop-on cover that can be removed for augering (see pages 202–203) when clogs occur. If your drain has a cover like this, snap it in place now.

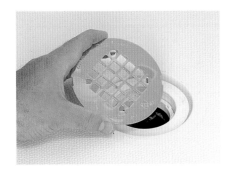

## 17 INSTALL COVER PLATES AND HANDLES.

Next, install the cover plates or escutcheons that cover the holes in the side panel for the handle or handles. Then add the handle(s) and secure with the screws provided.

## 18 INSTALL THE SHOWERHEAD.

To complete the shower install, attach the showerhead to the shower arm, and the shower arm to the shower elbow. See pages 142–149 for step-by-step directions on installing a shower faucet.

# Installing a Shower Door

If you're tired of grappling with a shower curtain and would like to step up to a shower door, you'll find a wide variety of units available. Options include finish (chrome and brass are common), along with glass type: clear, opaque, and patterned. Although models vary from one manufacturer to the next, most sliding doors (often called bypass doors) consist of six main parts: a top and bottom track, a pair of side channels, and two doors (drawing below). The doors are suspended from, and slide along, the top track that bridges the side channels. Side channels fit into the bottom track and when caulked, create a watertight seal around the perimeter.

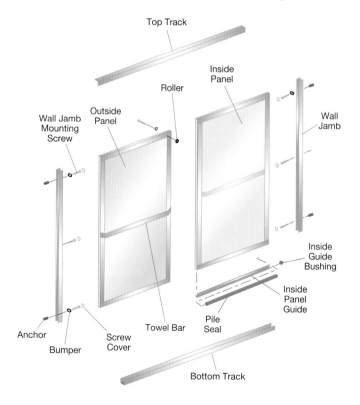

Top Track

Inside Panel

Roller

Outside Panel

Wall Jamb Mounting Screw

Wall Jamb

Inside Guide Bushing

Inside Panel Guide

Anchor

Bumper

Screw Cover

Towel Bar

Pile Seal

Bottom Track

Channels in the bottom track keep the doors from hitting each other as they're slid from side to side. One or more towel racks may attach to the doors. Note: Most door kits are designed to accommodate a range of shower sizes—make sure to have your shower measurements in hand when shopping for one.

**1 MEASURE AND CUT THE BOTTOM TRACK.** Measure the width of your shower receptor. Follow the manufacturer's directions to cut the track to length if necessary (usually it's cut narrower than the opening to allow the side channels to slip over its ends).

**2 LOCATE THE BOTTOM TRACK.** Once you've cut the track to length, position it on the ledge of the shower receptor so there's an even gap at both ends and so the track is centered on the width of the receptor's ledge. Use masking tape to temporarily hold it in place (as shown here).

**3 MARK THE SIDE CHANNELS.** Since the side channels support the top track and therefore the weight of the doors, it's important that they be firmly secured to the walls. In most situations they won't align with wall studs. That means you'll need to secure them with hollow wall anchors. Use a level to plumb each side channel, and mark the mounting hole locations onto the walls.

**4** DRILL HOLES FOR THE CHANNEL FASTENERS.
Drill appropriate-sized holes in the side panels for your hollow anchors (usually enclosed with your shower door kit).

**5** INSTALL THE SIDE CHANNEL FASTENERS.
Once the anchor holes are drilled in the side panels, insert the hollow wall fasteners. Take care to tap these in place gently to keep from damaging the side panels.

P R O  T I P

# Drilling through Tile

If the walls of your shower are tiled, you'll need a glass bit or tile bit to drill through the tile. Glass or tile bits are designed just for "drilling" (actually grinding) through tile and glass. They use a diamond-pointed or a tungsten-carbide tip to grind their way through the glass. You can find glass and tile bits at hardware stores and home centers. When using one, apply constant and steady pressure; if you see visible chips, you're not pressing hard or steadily enough.

**6** **INSTALL THE BOTTOM TRACK.** Before you install the side channels, the bottom track needs to be removed, caulked, and replaced. Apply beads of silicone as directed and set the track in place. Some instructions will advise using masking or duct tape to hold the track in place until the silicone sets up.

**7** **SECURE THE SIDE CHANNELS.** Now you can reposition each of the side channels and secure them to the wall using the screws provided with the hollow wall anchors. Note: It's a good idea to use a level to check for plumb as you do this and to adjust the position of the side channel(s) as needed.

**8** **INSTALL THE TOP TRACK.** With the side channels in place, you can attach the top track, cutting it if necessary to fit. Note: As the top track usually fits over the side channels, it's cut longer than the bottom track. Set the track in place over the side channels, and secure it to the side channels if screws are provided for this. On some door kits, the weight of the doors is all that's required to hold the top track in place.

**9** **ATTACH THE DOOR ROLLERS.** Now you can install the doors. You'll probably first have to attach the rollers to the top flanges of each door. Take care to follow the directions, as the rollers are often installed differently on each door.

# 10 INSTALL THE DOOR PANELS.

When the rollers are in place, grip a door firmly on the sides with both hands and lift it up into a channel in the top track so the rollers slip in place. Then pivot the door in so the bottom rests in the corresponding channel in the bottom track. Repeat for the other door. If towel racks are supplied, attach them now.

# 11 FINAL CAULKING.

All that's left is to follow the manufacturer's directions on applying caulk to create a watertight seal between the side channels and bottom track and the shower enclosure. Make sure to use 100% silicone for this: It will hold up the best over time.

# Installing a Water Filter

Installing some form of water treatment device is becoming a popular plumbing project. Why? The presence of more VOCs (volatile organic chemicals) in local water supplies, such as pesticides or herbicides, and other contaminants like lead. There is a wide variety of treatment devices available, ranging from simple in-line cartridges to reverse-osmosis systems with under-sink storage tanks. In between these are easy-to-install dual-cartridge systems like the one shown here. This type of system requires no electrical power, hooks up directly to your cold water line, and installs in an afternoon. It's important to note that cartridge-based systems require you to periodically change the cartridge. When shopping for a water-treatment device, look for one that has a built-in indicator that tells you when it's time to replace the filters.

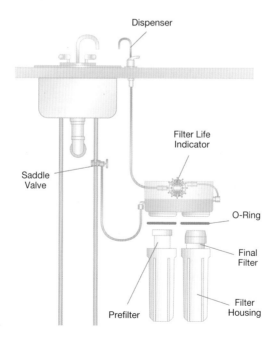

Dispenser

Filter Life Indicator

Saddle Valve

O-Ring

Final Filter

Filter Housing

Prefilter

**1 INSTALL THE DIS-PENSER.** The first step in installing a water filter is to mount the dispenser. Most of these are designed to fit in the extra hole in a sink top. If this hole is used by your sprayer, you'll need to drill a hole for the dispenser (or switch to a pull-out faucet sprayer; see pages 138–141). Follow the manufacturer's directions for the location and the size of the hole.

**2 DRILL A HOLE FOR THE SADDLE VALVE.** Most water filters tap into the existing cold water supply line via a saddle valve. To install a saddle valve, first turn off the water supply and open the faucet to drain the line. Then, following the manufacturer's directions, drill a small hole in the supply line at the specified location.

INSTALLING FIXTURES

**3** **INSTALL THE SADDLE VALVE.** Next, turn the handle on the valve to expose the lance and position the valve over the pipe so the lance fits in the hole. Attach the back plate of the valve and tighten the nuts to lock it in place.

**4** **MOUNT THE FILTER.** Position the filter roughly between the cold water line and the dispenser. Make sure to leave the specified clearance between the bottom of the filter and the cabinet bottom to allow for cartridge removal and replacement. Secure the filter to the cabinet back or wall with the screws provided.

**5** **CONNECT TO THE SADDLE VALVE.** The filter is hooked up to the water dispenser and the saddle valve with tubing and compression fittings. Start by cutting a length of tubing to reach from the saddle valve to the filter. Press the tubing into the compression fitting and thread this onto the saddle valve; tighten with an adjustable wrench.

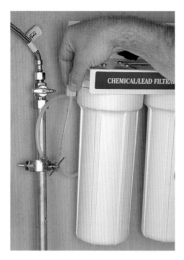

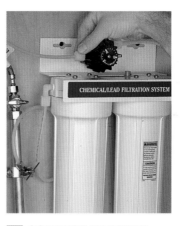

## 6 CONNECT TO THE FILTER.
Next, insert the opposite end of the tubing into another compression fitting and thread this onto the inlet port of the system. Tighten the nut until it's hand-tight, and then use an adjustable wrench to tighten the nut another turn or turn-and-a-half.

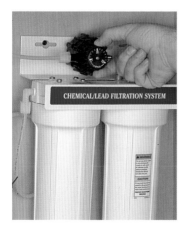

## 7 CONNECT FILTERED WATER TO THE DIS-PENSER.
Finally, cut a piece of tubing to run from the outlet port of the system to the water dispenser. Insert compression fittings on both ends, and thread the nuts onto the dispenser and the system.

## 8 ADJUST THE FILTER INDICATOR.
Turn on the water supply and open the water dispenser; let it run for about 5 minutes to flush any air or carbon particles out of the system. Most manufacturers recommend that you let the water run for about 20 seconds each time before using the water for drinking or cooking. If your filter system has a built-in indictor for cartridge replace-ment, follow the directions for setting the indicator knob to its starting position.

# 8

# Troubleshooting

**H**EY, ISN'T IT GREAT how well that toilet/shower/ sink is working? This thought rarely occurs to us: We ignore fixtures when they do their jobs properly—but let water flow where it shouldn't, and it's an instant attention-getter. When plumbing emergencies do occur, they're usually either clogs or leaks. In this chapter we'll show you how to handle common clogs in bathroom sinks, kitchen sinks, shower and tub drains, and finally, toilets. We'll also cover techniques for stopping and preventing leaks in pipes, sinks, faucets, and toilets.

# Emac... 

# Emergency Repairs

Although a clog can create a messy problem—a toilet or sink overflows—a leak can pose a real emergency: water spewing out of a wall, floor, or ceiling. In these instances, it's vital that everyone in your home knows not only where the main water shut-off is, but also how to stop the flow of water.

If you do have a leak, the first task is to find its source. It some cases, it's apparent: The water is pouring out of a pipe that burst. At other times you'll have to do some detective work to track it down. Start by looking for obvious clues like a picture newly installed on a wall with a large nail, or an overflowing tub or toilet. If it isn't obvious, try tracing the water flow back to the nearest fixture. Note: The methods described here are *temporary* repairs to keep water flowing in your home until permanent repairs can be made.

**PENCIL TIP.** You can temporarily stem the flow of water in a small pinhole leak by inserting the tip of a pencil into the hole and snapping it off. (Make sure to turn off the water first and dry the pipe.) To keep the tip in place, wrap a couple layers of electrician's tape or duct tape around the hole.

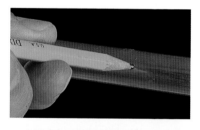

**FAUCET WASHER.** If you've got a small sheet-metal screw and a faucet washer rolling around in your junk drawer, they can be pressed into service to temporarily stop a small leak. Just make sure the screw is short enough so it won't puncture the opposite side of the pipe.

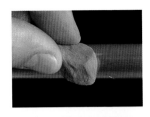

**EPOXY PUTTY.** You'll find epoxy putty at virtually every hardware store and plumbing supply house in the world; it's easy to use and waterproof. Most versions come in a Tootsie Roll–like log of two differently colored inner and outer layers. When you cut off a piece and massage the two parts together, the epoxy is activated. Apply it to the leak and let it set up before turning the water back on.

**METAL REPAIR CLAMP.** Another repair option is to wrap the pipe with scraps of rubber held in place with hose clamps. Pipe repair kits (like the one shown here) are commonly available; they're just two squares of flexible rubber that are squeezed tightly against the pipe with a pair of metal clamps.

---

**Q U I C K   F I X**

# Frozen Pipe

🚫 Pipe leaks can be caused by water freezing in the pipe. Ice can expand and burst the pipe or force apart a sweated joint. Regardless of the damage, the first step is to turn off the water and thaw the frozen pipe. You can do this with a heat gun or propane torch (as shown), or wrap the pipe with rags and pour hot water on it. If the pipe is burst, you can patch it with a pipe repair kit (see above), or replace the damaged section with new pipe. For joints where the ice caused the joint to fail, you can usually just re-sweat the joint (see page 65).

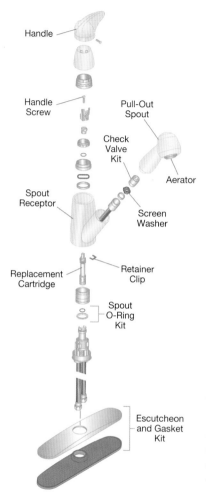

Handle

Handle
Screw

Pull-Out
Spout

Check
Valve
Kit

Aerator

Spout
Receptor

Screen
Washer

Replacement
Cartridge

Retainer
Clip

Spout
O-Ring
Kit

Escutcheon
and Gasket
Kit

# Faucet Leaks

When the "drip-drip" drives you crazy, don't reach for earplugs or for the phone to call in a pro. You can fix it yourself. Most faucets are one of two types: cartridge or compression.

## Cartridge faucets

On a cartridge faucet (drawing at left), the cartridge moves up and down within the body as the handle is adjusted to control the amount and temperature of the water to the spout.

**1 REMOVE THE HANDLE.**
Before you begin work on a faucet, shut off the water and open the faucet to drain the lines. Then remove the handle. It may be attached via a setscrew (as shown here), which is often concealed by trim. Alternatively, a cap can conceal a screw that secures the handle to the body.

**2 REMOVE THE RETAINING CLIP.**
On many faucets, the cartridge is held in place with a retaining clip. Use a pair of needle-nose pliers to grip the clip and pull it out.

**3** **REMOVE THE CARTRIDGE.** The old cartridge can now be pulled up and out of the faucet body. Quite often, as you pull you'll need to wiggle the cartridge from side to side to free it from the socket body. If you're repairing the cartridge and not replacing it, wrap tape around the pliers' jaws to keep from scratching the stem. To install a new cartridge, just reverse the disassembly process. Turn on the water and test. If hot and cold are reversed, turn off the water, disassemble, and rotate the stem one half-turn.

**4** **REPLACING O-RINGS.** In many cases, you can stop a cartridge faucet from leaking by replacing its O-rings. Most home centers and hardware stores sell repair kits just for this; see the sidebar below. Pry off the old O-rings and lubri- cate the new O-rings with petroleum jelly (or the lubricant supplied in the repair kit) and roll them into the correct grooves.

## QUICK FIX

# Cartridge Repair Kits

Cartridge repair kits contain the parts of a cartridge that are exposed to the bulk of the wear and tear: the O-rings and springs. Repair kits frequently come with a small tube of O-ring lubricant that helps keep the rings flexible over time. Note that these kits are made for specific brands and models, so be sure you have the make and model number of your faucet when shopping for a repair kit.

## Compression faucets

Compression faucets (often called seat-and-washer faucets) individually control the flow of either hot or cold water that is then sent to the spout to mix. As the handle is rotated, a valve stem rises or lowers to allow more or less water through to the spout. A rubber washer on the end of the stem presses against the seat in the base of the faucet body to stop the flow when the handle is turned off.

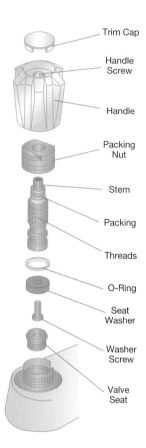

Trim Cap

Handle Screw

Handle

Packing Nut

Stem

Packing

Threads

O-Ring

Seat Washer

Washer Screw

Valve Seat

**1 REMOVE THE HANDLE.** If water drips out from the spout when the handles are off, the problem is most likely the seat washer. If it leaks from the handle, it's the O-rings. To repair the faucet in either case, start by turning off water at the shut-off valves. Then remove the handle screws and lift off the handles.

**2 REMOVE THE VALVE STEM.** Next, use a pair of channel-type pliers or an adjustable wrench to loosen the packing nut holding the stem in place. Then lift the valve stem out of the faucet body.

**3** **REMOVE THE OLD WASHER.** To replace the stem washer, hold the stem in one hand and use a screwdriver to loosen the washer retaining screw. In some cases, it may help to reinstall the handle on the stem for better leverage. If the screw is stubborn, apply a

few drops of penetrating oil and let it sit for 15 minutes before trying again. Remove the screw and pry out the old washer.

**4** **INSTALL THE NEW WASHER.** Replace the washer with a direct replacement part—you can find most types at a hardware store or a plumbing supply house. Position the new washer and install the retaining screw. Reassemble the faucet, turn the water back on, and test. If the faucet still drips with the handle off, you may need to replace the valve seats; see the sidebar below.

**P R O   T I P**

# Replacing a Valve Seat

If replacing a seat washer doesn't stop a faucet from dripping, the problem is likely a damaged seat. To check for damage, turn off the water and remove the handle and stem. Insert a finger into the faucet body and gently rub your fingertip around the lip of the seat—it should be smooth and level. If you find burrs or scratches, replace the seat. Use a seat wrench (an L-shaped tool that has a different tip on each end to fit various seats); the L-handle provides the leverage you'll need to loosen a valve seat.

# Water Pressure Problems

Inadequate water pressure is a common homeowner complaint. The problem can be caused by a number of things, individually or combined. And there may be remedies to these problems, or not—or, the solution may be too expensive. For example, if the lack of pressure is caused by the local water company, there's often not much you can do about it except complain. Likewise, if the local water has plenty of pressure but it's insufficient in your home, the culprit may be water scale.

## Water scale

Over time, built-up deposits can effectively decrease the diameter of a pipe, as shown in the drawing at right. These deposits—often referred to as scale—attach themselves to the interior walls of the pipe, much like the way cholesterol can line human arteries. If the water pressure is fine at the street but not at your outside hose, the problem is probably the line running from the street to your home. This can be replaced, but it requires digging a trench, relaying the line, and recovering. The cost and the mess can be enormous.

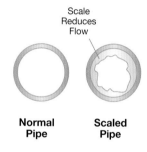

Scale Reduces Flow

**Normal Pipe**  **Scaled Pipe**

If the pressure is fine as it enters your dwelling but not elsewhere in the home, the problem could be scale built up in interior piping. Or, the cause could be clogged aerators and screens; see the opposite page. If scale is a problem within your home, consider installing a whole-house water treatment device to reduce the chances of continued scale problems.

## Aerators and screens

Insufficient water pressure at a fixture can be caused by aerators or screens that fit over the ends of faucets or inside faucets and showerheads to trap debris. Over time, these will clog (particularly if you have hard water), reducing water flow. The solution can be as simple as cleaning. Remove the aerator or screen and soak it in vinegar for an hour or so to loosen up stubborn sediment. Then scrub it with a toothbrush. If the sediment is hardened and can't be removed with cleaning, you'll need to buy a replacement.

**FAUCET AERATORS.** A clogged aerator not only reduces water flow, but it can also cause the water that does come out to spray in a wild or erratic pattern. Sometimes you can unscrew an aerator by hand; often you'll need an adjustable-jaw wrench to break it free from the spout. Be especially careful as
you thread a cleaned or replacement aerator back on the spout: The threads are very fine and are easily cross-threaded.

**SHOWERHEADS.** Sandwiched in between the showerhead and the shower arm on many showers is a debris screen. These can be cleaned (as described above) or replaced if needed.

**PULL-OUT SPRAYER FAUCETS.** The filter screen on many pull-out sprayer faucets frequently is located between the faucet handle and the pull-out sprayer hose (as shown here). Alternatively, they can be found where the sprayer hose connects to the faucet body.

TROUBLESHOOTING

# Toilet Clogs

Clogs in toilets primarily occur in the internal trap and are caused by solid waste, facial tissue, paper towels, cloth diapers, and sanitary napkins. If you know the clog is a cloth diaper or a sanitary napkin, your best bet is to snag it with a closet auger (see page 191) and pull it out. Don't try to force it down the drain with a plunger: It'll most likely get stuck in the main waste stack, and you'll have a bigger problem. Other clog locations are in the closet bend, at the junction of the sanitary tee, and in the main waste stack, as illustrated in the drawing at right.

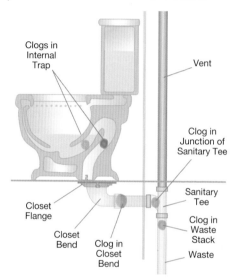

Clogs in Internal Trap

Vent

Clog in Junction of Sanitary Tee

Sanitary Tee

Closet Flange

Clog in Waste Stack

Closet Bend

Clog in Closet Bend

Waste

## Ball-Type Plungers

A ball-type plunger with a skirt or flange works best for plunging toilets because it creates a better seal. Pull the flexible skirt out when working on a toilet or press it up into the ball (as shown here) when plunging flat-bottomed fixtures like a kitchen sink, a shower, or a bathtub.

**PLUNGE, PLUNGE, AND THEN PLUNGE SOME MORE.** The most common mistake that most people make when plunging is not using enough force and not plunging long enough. To plunge a toilet, start by making sure there's sufficient water in the bowl to cover the rubber cup and create a seal. Then plunge with vigor. You're going to splash some, so have towels handy. Rest between spurts of plunging, and make sure to maintain the water level in the bowl to cover the cup. Repeat six to eight times.

**USING A CLOSET AUGER.** If plunging doesn't get the job done, try a closet auger (page 54). Turn the handle of the auger as you insert the end into the trap opening. As you press the end into the opening, continue to turn the handle. A series of turns will most likely clear an obstruction.

**AUGERING THE WASTE LINE.** If the clog isn't in the toilet, it's in the closet bend, sanitary tee, or main waste stack. All of these will require augering the line with a snake or hand-powered auger (pages 54–55). The first thing to do is to remove the toilet, as described on pages 86–89. Then insert the end of the snake or hand-powered auger into the closet bend and begin rotating the snake or auger so its end will bore through the clog. If you still can't clear the clog, you'll need to call in a pro.

# Toilet Leaks

There are four main areas for potential leaks on a toilet. From the ground up, they are: the seal between the bowl and closet flange, the water supply and shut-off area, the ballcock connection to the tank, and the tank-to-bowl connection, as shown in the drawing below.

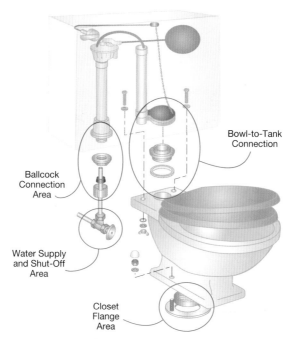

Bowl-to-Tank Connection

Ballcock Connection Area

Water Supply and Shut-Off Area

Closet Flange Area

**REPLACE THE WAX RING.** If there's water seeping out from under your toilet, chances are that the wax ring needs replacing. Before you do this, try tightening the mounting bolts (see page 155). If this doesn't stop the leak, it's time for a new ring. There are four basic steps to replacing a wax ring: removing the toilet and old wax ring, setting the bowl on the new wax ring, connecting the tank, and reconnecting the water supply (see pages 86–89 and 152–159).

**TIGHTEN THE TANK-TO-BOWL CONNECTION.** If you see water dripping down the back of the bowl, the problem may be a poor seal between the tank and the bowl. Before replacing the spud washer (see below), try tightening the tank bolts. Insert the tip of a screwdriver into the threaded head of one of the tank bolts. Hold the bolt still with the screwdriver, and tighten the tank bolt with a socket wrench or an adjustable wrench.

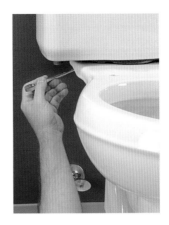

**REPLACE THE SPUD WASHER.** The seal between the tank and bowl is created by the spud washer. Over time, these can deteriorate and

fail. Replacements can be found at most plumbing supply houses and occasionally at a hardware store or home center. Remove the tank and flip it upside down. Pry off the old spud washer, and install the new one. Replace the tank and test.

**CHECK THE SUPPLY LINE CONNECTIONS.** Leaks can also be caused by the connections between the shut-off valve and the ballcock assembly, or by the shut-off valve itself—it all depends on where the water is leaking from. If it's coming out of the shut-off valve, tighten the connection or reapply Teflon tape to the valve's threads. If the leak is under the tank where the ballcock connects to the tank, tighten this or replace the gasket.

# Toilet Flush Problems

The automatic valve inside a toilet tank that refills the tank after a flush is called a ballcock, or fill valve. Although it looks complicated (see below), it's just a valve—a cup connected to the valve moves along with the water level in the tank to turn the water on or off.

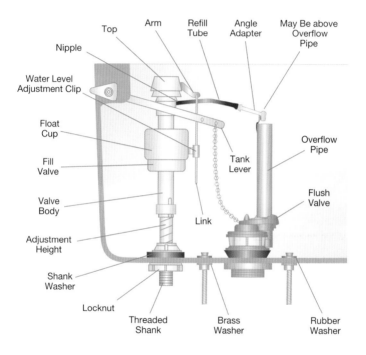

**CHECK THE OVERFLOW PIPE.** If a toilet runs constantly, water in the tank may be draining over and down into the overflow pipe. Properly set, the water should be ½" to 1" *below* the top of the pipe. To adjust a float-arm ballcock, lower the water level by bending the float arm down; bend it up to raise the level. On a float cup, raise or lower the cup (and water level) by pinching the metal clip that fits over a pull rod. Then slide the clip (and cup) up or down. Some ballcocks use a metered fill valve to control the water level. These can be adjusted by turning an adjustment screw on top of the valve.

**REPLACING A FLAPPER.** A constantly running toilet can be caused by a faulty flapper. The flapper allows tank water to flow into the bowl. This in turn causes the cup to lower and refill the tank. To replace a flapper, first remove the chain from the lift arm, and then slip each ear of the flapper off its associated lug on the sides of the overflow tube (top photo).

**REPLACE THE FILL VALVE.** To replace a fill valve, turn off water to the toilet and empty the tank. Disconnect the refill tube (middle photo) and remove the supply line running to the valve. Remove the washer that secures the valve to the tank and lift out the valve. Reverse these steps to install the new fill valve.

**CHECK THE CHAIN.** If a lift chain is too short, the flapper won't form a seal and water will leak into the bowl, causing the fill valve to refill the tank. Try moving the chain to another hole in the trip lever (photo at right). If this doesn't work, remove or add links to the chain.

QUICK FIX

# Toilet Repair Kits

When the inner workings of a toilet tank fail, chances are that all of the parts are worn out. That's why repair parts makers sell kits with all you'll need to rebuild a toilet. Parts include a flush valve, flapper valve, overflow pipe, and spud washer.

# Sink Clogs

**H**ow you deal with a clogged sink depends on whether it's a bathroom sink or kitchen sink.

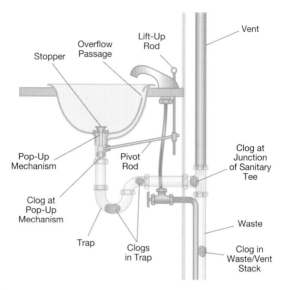

## Bathroom sinks

Bathroom sinks differ from other sinks (such as kitchen and utility sinks) in that they have a built-in system to handle overflows. Slots or holes near the top of the basin allow water that reaches this preset level to flow through an overflow passage into the drain (drawing above). Another unique feature of bathroom sinks is that the drain stoppers are usually connected to a pop-up mechanism that allows the stopper to be raised or lowered remotely. While handy, this unfortunately creates an area in the drainpipe (where the linkage connects to the stopper) that is prone to clogging. Clogs in a bathroom sink are usually caused by a buildup of soap, hair, toothpaste residue, etc. Although the knee-jerk reaction to a clog is to use a chemical drain cleaner, don't (see page 55).

# 1 REMOVE THE STOPPER.
Before you can effectively plunge a bathroom sink, you'll need to remove the drain stopper. There are two versions of drain stoppers: One type simply lifts directly out; with the other type you'll have to remove a pivot rod that connects to the pop-up mechanism (see pages 130–133). Clean the drain stopper and set it aside.

# 2 BLOCK THE OVERFLOW AND PLUNGE. Next,
block off the overflow to prevent water from bouncing off the clog and traveling up the overflow. To block the overflow, insert a damp rag or a sponge. To plunge, keep the water in the sink above the cup; plunge with a series of quick, strong strokes. Repeat as needed to clear the clog.

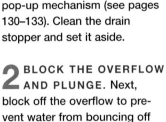

# 3 REMOVE AND CLEAN THE TRAP.
If plunging didn't work, remove and clean the trap. Place a bucket under the trap and have towels handy for spills. Loosen the slip nuts on both ends of the trap and remove it. Shake the trap over the bucket to dislodge debris. If the clog is compacted in the trap, use a snake or auger to clear the trap.

# 4 AUGER THE WASTE LINE. If you
pull off the trap and discover there's no debris, the clog is most likely in the line leading to the sanitary tee, which connects to the waste stack. Insert the end of a snake into the drain opening until you feel resistance. Then rotate the snake to clear the clog.

TROUBLESHOOTING

## Kitchen sinks

Most kitchen sinks have at least two basins that are inter-connected via a T- or Y-shaped connector. To further complicate things, a garbage disposer is often attached to one basin, as shown in the drawing below.

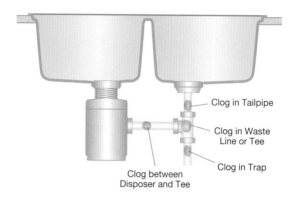

Clog in Tailpipe

Clog in Waste
Line or Tee

Clog in Trap

Clog between
Disposer and Tee

**1** **BLOCK THE OTHER BASIN AND PLUNGE.** If your kitchen sink has two basins, you'll need to seal off one side before you can plunge. If you don't, the water that the plunger forces down the drain will come up into the other basin instead of working to break the clog free. The simplest way to seal it is to use the strainer. In order to use both hands on the plunger, you'll need a helper to hold the strainer in place. If your sink has a garbage disposer built in and it uses an air gap, you'll need to seal off the hose going to the air gap. If you don't, the water you force down the drain will have an alternate, easier path to follow instead of forcing out the clog. Use a small C-clamp or a pair of vise grips to temporarily pinch the walls of the tubing together. Fill the basin with water so it's over the rubber cup of the plunger. Give the plunger a series of hard downward strokes and lift it sharply out. Repeat as needed to clear the clog.

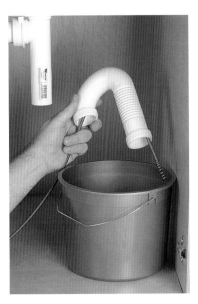

## 2 CLEAR THE TRAP.

If both basins are still clogged, and plunging doesn't clear out the obstruction, the next step is to remove the trap fittings and clear out any debris. Start by placing a bucket under the trap and have some rags handy. Loosen the slip nuts that hold the trap in place and remove it. Clear out any debris with a snake or auger.

## 3 AUGER THE WASTE LINE.

If there isn't any obstruction in the trap, the clog is either in the drain line to the sanitary tee, or in the waste stack itself. In either case, the solution is to clear the clog with a hand auger. Push the end of the auger into the line until you feel resistance. Then lock the auger in place with the thumbscrew on the end of the auger. Rotate the auger's handle to force the auger tip to bore through the clog. Once the clog is cleared, reassemble the drain line and flush it with water.

TROUBLESHOOTING

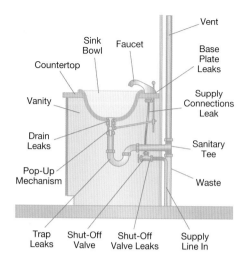

Countertop
Sink Bowl
Faucet
Vent
Base Plate Leaks
Supply Connections Leak
Vanity
Sanitary Tee
Drain Leaks
Pop-Up Mechanism
Waste
Trap Leaks
Shut-Off Valve
Shut-Off Valve Leaks
Supply Line In

# Sink Leaks

Regardless of whether a sink is installed in a bathroom or in a kitchen, sinks have much in common. Both have lines running up to the faucet to supply hot and cold water and are controlled by shut-off valves. Both have a trap on the waste line to block sewer gas, and both use some sort of stopper to close the drain so the basin(s) can be filled (drawing above). So the problems you'll encounter have similar solutions. Leaks can often be eliminated with plumber's putty or by tightening a nut on a drain body or trap, or the clamp on a hose.

**SINK-TO-COUNTERTOP SEAL.** If you find water under your sink and it's not from the supply or waste lines, water may be seeping under the sink's rim. To remedy this, remove the sink (pages 82–83) and apply fresh plumber's putty or silicone caulk.

**SUPPLY-TO-VALVE CONNECTIONS.** Sink leaks can also be caused by the connections between the shut-off valves and the faucet, or by the shut-off valve itself. If water is coming out of the valve, tighten the connection or reapply Teflon tape to the valves' threads. If it's coming from the supply line–to-faucet connection, remove the line, apply Teflon tape, and reinstall.

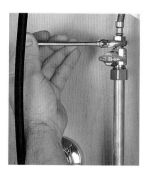

**TRAP AND WASTE LINE LEAKS.**
Another source of leaks under the sink is a tailpiece, trap, or other waste line that has worked loose—often the result of being bumped by objects placed under the sink. To stop a leak, tighten the compression fittings by hand until they're snug and then tighten them another quarter-turn. If the leak persists, replace the tailpiece compression ring; see the sidebar below.

**STRAINER/DRAIN LEAKS.**
If water is leaking out from around the base of the sink flange, you'll need to either install fresh putty around the flange, or if it's cracked, replace it. Start by removing the trap and drain body. Push the flange out of the sink from underneath and remove the old putty. If the flange looks good, apply a coil of plumber's putty to the flange lip and reinstall it. Replace a broken or deformed flange.

---

Q U I C K  F I X

## Plastic Compression Washers

Most waste line and trap connections are held together via compression fittings. With time, the plastic washers that actually compress to create the seal can become stiff and lose their ability to form a good seal. That's why it's a good idea to keep an assortment of plastic compression washers on hand to stop waste line leaks that can't be repaired just by tightening the fitting.

TROUBLESHOOTING

# Bathtub Clogs

Clumps of hair and soap sludge are the most common culprits in shower and bathtub clogs. Shower drains are fairly simple to clear, since you just remove the drain cap and then auger the line. Bathtubs, on the other hand, present a real challenge because of the mechanism used to stop up the drain when filling a tub.

There are two basic types of these mechanisms: One uses a stopper and is remotely activated via a trip lever, much like the pop-up mechanism of a typical bathroom sink. The other type uses a rod-and-bucket assembly to stop water from draining out of the tub. The rod-and-bucket type is the most common. A two-piece adjustable rod that fits inside the tailpiece attaches to a trip lever via a clevis pin on one end, and supports the bucket on the other end (drawing above). When the trip lever is activated, it lowers the bucket into the tee, blocking the flow of water coming from the waste elbow. It's probably no surprise that this rod-and-bucket assembly has a well-deserved reputation for catching hair and soap to form quick clogs. This is why the first step in unclogging a bathtub is to remove this assembly; see page 203.

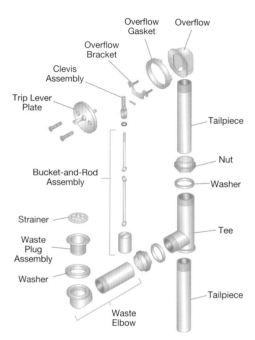

Overflow Gasket
Overflow
Overflow Bracket
Clevis Assembly
Trip Lever Plate
Tailpiece
Nut
Bucket-and-Rod Assembly
Washer
Strainer
Tee
Waste Plug Assembly
Washer
Tailpiece
Waste Elbow

**1** **REMOVE THE TRIP LEVER.** To remove the bucket-and-rod assembly, first remove the trip lever by unscrewing the screws that hold the trip lever to the overflow.

**2** **PULL OUT THE LIFT ROD AND BUCKET.** Then pull out the bucket-and-rod assembly. Quite often you'll discover a hair ball wrapped around the bucket. If the water flowed down the drain when you pulled that out, you're in luck. If removing the waste and overflow linkage didn't clear the clog, it's time to pull out the plunger. Make sure there's sufficient water in the tub or shower to completely cover the rubber cup of the plunger. Block off the overflow tube with a damp rag, and plunge with a series of sharp, hard strokes. Continue until the drain is cleared.

**3** **AUGER THE WASTE LINE.** If plunging didn't work, you'll need to auger the overflow and drain. Start by augering the waste elbow. Then auger the waste line by passing your snake or hand-powered auger down through the tailpiece and past the sanitary tee until you strike and clear the clog.

TROUBLESHOOTING

# Branch and Main Line Clogs

If plunging and augering a drain line doesn't clear a clog, the clog is located either in a branch or main waste line, or in the sewer service line. You can often clear clogs in branch and main lines by systematically working from the clogged drain line toward the service sewer line. If you're lucky, your home will have a cleanout plug at the end of each branch line. This is the first place to start. If you still don't locate the clog, advance to the main waste stack. If your problem is in the sewer service line, we suggest calling in a professional drain-service company.

The main waste/vent stack transports solid and liquid waste out of your home. The system uses gravity to move the wastewater from sinks, showers, tubs, and toilets out of the fixtures and into a waste line. This waste line empties into the municipal sewer or a private septic tank. Depending on the number of fixtures and layout of your home, you may have a single vent, or multiple vents (drawing at left). The single or larger line is the main waste/vent stack, or the soil stack. Branch drains connect to the main stack via fittings called sanitary tees. These fittings provide an angled junction that allows waste to flow freely.

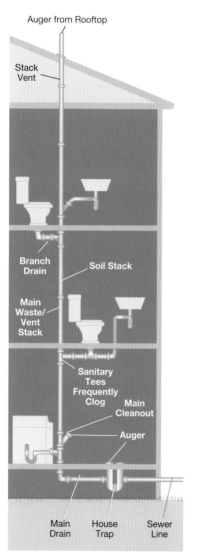

Auger from Rooftop

Stack Vent

Branch Drain

Soil Stack

Main Waste/ Vent Stack

Sanitary Tees Frequently Clog

Main Cleanout

Auger

Main Drain

House Trap

Sewer Line

## Augering a branch or main line

If you are fortunate enough to have cleanout plugs in the ends of your branch lines, locate the one nearest the clogged fixture (they're often located between the floor joists in a basement).

**1 REMOVE THE PLUG.**
Before loosening the plug, place a bucket underneath the opening and have plenty of towels or rags on hand. As you unscrew the cleanout plug, don't stand directly beneath or in front of the opening. Proceed with caution: The weight of the backed-up water can make it shoot out of the opening like a geyser.

**2 AUGER THE LINE.** If you removed the plug and there was little water, the clog is likely between the cleanout and the sanitary tee closest to the clogged fixture. Try clearing this with a hand auger. Feed the auger in until you feel resistance. Lock it and crank the arm. Continue until the clog is cleared. Alternatively, if have access to a long snake, you can clear the stack by augering down through the vent stack on your roof (use caution when working with ladders and on a roof).

## USING A SEWER SERVICE

Plunging didn't work. Augering the drain and branch lines didn't work. You've got a major clog in either the main waste stack or the sewer line. You can try to auger the line yourself, but this is best left to a professional sewer service, with their power augers. Yes, you can rent a power auger, but these powerful machines require an experienced touch—that's because they are capable of boring right through the walls of your drainpipes if not used properly.

# Index

## A
ABS pipe, 21, 22
Aerators, 129, 189
Augers
 described, 54, 55
 using, 191, 205

## B
Ball valves, 26
Bathrooms, 14. *See also specific fixtures*
Bathtubs
 clogs in, 202–203
 rough-in for, 75
 types of, 38, 41

## C
Cement, for plastic pipe, 50, 68
Clogs
 bathtub, 202–203
 branch and main line, 204–205
 chemical removers for, 55
 sink, 196–199
 toilet, 190–191
 tools for, 55–56
Closet flange bolts, 88, 154
Compression fittings, 73, 201
Copper pipe, 20, 48–49, 62–65
CPVC pipe, 21. *See also* Plastic pipe

## D
Deburring tools, 50, 67
Drill bits, for tile, 172

Drills, 53
DWV (drain-waste-vent) pipe, 21

## F
Faucets
 filter screens on, 189
 installing in showers, 142–149
 installing in sinks, 126–129, 134–141
 leaks in, 184–187
 removing, 84–85
 repair kits for, 185
 types of, 34–36
Fittings, 20. *See also specific types of pipe*
Flexible lines
 described, 23, 70–71
 on garbage disposers, 98
 working with, 69
Flush problems, 194–195
Flux, 49

## G
Galvanized pipe, 22, 72
Garbage disposers, 98, 99
Gate valves, 26
Graphite-impregnated cord, 25

## H
Hacksaws, 46, 62, 66
Hammers, 53
Hanger bolts, installing, 113
Heat shields, 49
Hole saws, 51, 164

**J**

Johnny bolts, 88, 154

**K**

Kitchens, 15. *See also* Faucets;
Sinks

**L**

Layout tools, 52
Leaks
    faucet, 184–187
    pipe, 182–183
    sink, 200–201
    toilet, 192–193

**M**

Miter saws, 50

**P**

Pipe
    access space for, 76
    frozen, 183
    leaks in, 182–183
    measuring, 18–19
    securing, 77
    tools for, 46–50
    types of, 18–23
    watertight joints in, 78–81
Pipe joint compound, 24, 72, 80
Plaster guards, 147
Plastic pipe, 21, 50, 66–68
Plastic-tubing pliers, 47, 66
Plumber's putty, 24, 79
Plungers, 54, 190
Pop-up mechanisms, 130–133
Putty knives, 52
PVC. *See* Plastic pipe

**R**

Reamers, 58, 63
Reciprocating saws, 51
Repair kits, 185, 195
Rough-in plumbing, 74–76

**S**

Saddle valves, 27
Saws. *See specific types*

Screwdrivers, 52
Sealants, 24–25, 78–81
Sewer services, 205
Shower doors, 40, 170–175
Showerheads, 169, 189
Showers
    faucets for, 142–149
    installing, 160–169
    rough-in for, 75
    types of, 29, 41
Shut-off valves, 9, 90–91, 193
Sinks
    clogs in, 196–199
    installing in bathrooms, 100–117
    installing in kitchens, 118–123
    leaks in, 200–201
    materials for, 32–33
    openings for, 101, 118
    removing, 82–83
    rough-in for, 74
    supply lines for, 94–95
    types of, 28–31
    waste lines for, 96–99
Snakes, 55
Socket sets, 53
Soil stack, 10, 13
Spa showers, 41
Stud finders, 51
Supply lines
    connecting, 94–95, 141, 157
    described, 8, 23, 70
Sweating copper pipe, 64–65

**T**

Teflon tape, 25, 78
Toilets
    clogs in, 190–191
    flush problems in, 194–195
    installing, 152–159
    leaks in, 192–193
    removing, 86–89
    repair kits for, 195
    rough-in for, 75
    types of, 37
Tools. *See also specific tools*
    hand tools, 52–53
    pipe cutting and fitting, 46–50

Tools, *continued*
  remodel, 51
  specialty, 56–57
  waste-line, 54–55
  wrenches, 44–45, 56–57,
    60–61, 72, 84, 123
Torches and sparkers, 49
Traps, 11
Troubleshooting. *See* Clogs;
  Flush problems; Leaks;
  Water pressure
Tube-flaring tool, 57
Tubing benders, 69
Tubing cutters, 47, 62, 63, 69

**U**
Utility knives, 52

**V**
Valves, 9, 26–27, 136
Valve seats, replacing, 187

Vent system, 12–13

**W**
Waste lines
  clogs in, 204–205
  described, 10, 23, 71
  for sinks, 96–99
  tools for unclogging, 54–55
Water filters, 176–179
Water pressure, 188–189
Wax-free rings, 81
Wax rings
  described, 25, 81
  installing, 153
  removing, 89
  replacing, 192
Whirlpools, 41
Wire-brush pipe cleaners, 48
Wrenches
  types of, 44–45, 56–57, 123
  using, 60–61, 72, 84, 123

## Metric Equivalency Chart

Inches to millimeters and centimeters

| INCHES | MM | CM | INCHES | CM | INCHES | CM |
|---|---|---|---|---|---|---|
| 1/8 | 3 | 0.3 | 9 | 22.9 | 30 | 76.2 |
| 1/4 | 6 | 0.6 | 10 | 25.4 | 31 | 78.7 |
| 3/8 | 10 | 1.0 | 11 | 27.9 | 32 | 81.3 |
| 1/2 | 13 | 1.3 | 12 | 30.5 | 33 | 83.8 |
| 5/8 | 16 | 1.6 | 13 | 33.0 | 34 | 86.4 |
| 3/4 | 19 | 1.9 | 14 | 35.6 | 35 | 88.9 |
| 7/8 | 22 | 2.2 | 15 | 38.1 | 36 | 91.4 |
| 1 | 25 | 2.5 | 16 | 40.6 | 37 | 94.0 |
| 1 1/4 | 32 | 3.2 | 17 | 43.2 | 38 | 96.5 |
| 1 1/2 | 38 | 3.8 | 18 | 45.7 | 39 | 99.1 |
| 1 3/4 | 44 | 4.4 | 19 | 48.3 | 40 | 101.6 |
| 2 | 51 | 5.1 | 20 | 50.8 | 41 | 104.1 |
| 2 1/2 | 64 | 6.4 | 21 | 53.3 | 42 | 106.7 |
| 3 | 76 | 7.6 | 22 | 55.9 | 43 | 109.2 |
| 3 1/2 | 89 | 8.9 | 23 | 58.4 | 44 | 111.8 |
| 4 | 102 | 10.2 | 24 | 61.0 | 45 | 114.3 |
| 4 1/2 | 114 | 11.4 | 25 | 63.5 | 46 | 116.8 |
| 5 | 127 | 12.7 | 26 | 66.0 | 47 | 119.4 |
| 6 | 152 | 15.2 | 27 | 68.6 | 48 | 121.9 |
| 7 | 178 | 17.8 | 28 | 71.1 | 49 | 124.5 |
| 8 | 203 | 20.3 | 29 | 73.7 | 50 | 127.0 |

mm = millimeters    cm = centimeters

**PHOTO CREDITS**

**Bosch**
(www.boschtools.com):
page 50.

**Kohler**
(www.us.kohler.com):
page 28 (all), page 33 (top
and middle), page 34 (all),
page 35 (all), page 38,
page 39, page 41 (top).

**Moen**
(www.moen.com):
page 41 (bottom).

**Rocky Mountain
Hardware**
(www.rockymountain
hardware.com): page 33
(bottom).

**Sterling Plumbing**
(www.sterlingplumbing.
com): page 40 (all).